NATIONAL NUMERACY TESTS Y9

A+ National Practice Tests
graduated difficulty with solutions

Sue Ferguson
Sarah Hamper
Wendy Bodey

A+ National Numeracy Tests Year 9
1st Edition
Sue Ferguson
Sarah Hamper
Wendy Bodey

Publishing editors: Jane Moylan and Jana Raus
Editor: Kerry Nagle
Project packager: UC Publishing
Senior designers: Ami Sharpe & Vonda Pestana
Text designer: Ami Sharpe
Cover designer: Ami Sharpe
Cover image: Getty Images
Photo research: UC Publishing/Shelley Underwood
Production controller: Damian Almeida
Reprint: Katie McCappin
Typeset by UC Publishing Pty Ltd

Any URLs contained in this publication were checked for currency during the production process. Note, however, that the publisher cannot vouch for the ongoing currency of URLs.

Acknowledgements
VCE ® is a registered trademark of the VCAA. The VCAA does not endorse or make any warranties regarding this Cengage product. Current and past VCE Study Designs, VCE exams and related content can be accessed directly at www.vcaa.vic.edu.au

We would like to thank the following people for reviewing this title: Howard Reeves, Barbara D'Angelo and Col Stevenson.

The publisher would like to acknowledge the following permissions:

Shutterstock.com p17,80 Robert Spriggs; p28 The Darling, Hawkesbury Valley Homes; istockphotos.com, p47 MentalArt.

For product information and technology assistance,
in Australia call **1300 790 853**;
in New Zealand call **0800 449 725**

For permission to use material from this text or product, please email **aust.permissions@cengage.com**

ISBN 978 0 17 047067 4

Cengage Learning Australia
Level 7, 80 Dorcas Street
South Melbourne, Victoria Australia 3205

Cengage Learning New Zealand
Unit 4B Rosedale Office Park
331 Rosedale Road, Albany, North Shore 0632, NZ

For learning solutions, visit **cengage.com.au**

Printed in China by 1010 Printing International Limited.
1 2 3 4 5 6 7 26 25 24 23 22

Detailed Information

Introduction

▶ **To the student**

Literacy and numeracy are the fundamental building blocks of learning in any subject. Knowing what you can do and where you need to improve is vital for all Australian students, teachers and parents. The NAPLAN* (National Assessment Program – Literacy and Numeracy) tests in literacy and numeracy help governments find out how Australian students are progressing and help to identify what you know and what you don't know. The results of the tests also help you and your teachers plan what you need to learn next.

Tests can sometimes be a little daunting; however, there are practical steps you can take to ensure you will successfully sit a test. You will be best prepared for any test when you understand what is being tested, how you will be tested, and when you are mentally and physically prepared for the test. This book provides practical advice and strategies to ensure that you are test-ready and contains practice tests so you can see what to expect in the tests.

This book and the accompanying NelsonNet website provide:

✔ **Test tips:** advice on how to successfully sit tests, information about what is tested and how it is tested, hints on responding to different question types and how to act on your results.

✔ **Calculator skills:** provides a clear tutorial to ensure you make the best use of your calculator.

✔ **Comprehensive practice tests:** includes three non-calculator practice tests and three calculator-allowed practice tests. These tests are graduated in difficulty and content so that you can build your skills and confidence during the first part of the year leading up to the test.

✔ **Four full-length tests:** four full-length detachable tests (two non-calculator and two calculator-allowed tests), one set is for you to use as a practice test, and the other set is for you to hand in to your teacher. The full-length tests are of the same length and level of difficulty you can expect to find when you sit the NAPLAN* tests so are ideal for practise during the month leading up to the test.

✔ **Answers:** for easy reference, answers to all workbook questions can be found on NelsonNet, https://www.nelsonnet.com.au/free-resources.

✔ **Tips for teachers and handy checklists:** for easy reference teacher hints and checklists are provided on the NelsonNet website.

✔ **Useful icons:** icons are used throughout to help you quickly find other important information in the resources, support your understanding and clearly distinguish questions where you may use a calculator if you wish.

✔ You will also notice the following key features throughout this book:

Hot tips: keep one step ahead by heeding these hot tips. The tips often relate to things many students forget to do or can do better with a little planning.

World wide web link: provides a useful link to further information available on the Internet.

Calculator allowed: this indicates you may use a calculator to assist you to answer the question and when you check your work.

Non-calculator: this indicates you must complete any calculations without the use of a calculator.

About the authors

Sue Ferguson

Sue Ferguson has worked in mathematics and curriculum and assessment for many years. Presently Sue lectures at Victoria University, and prior to that she worked at the Australian Curriculum, Assessment and Reporting Authority (ACARA) heading a team of writers to develop the Australian Mathematics Curriculum. She has also worked at the Curriculum Corporation on The Le@rning Federation project where she provided mathematics advice on the development of digital curriculum resources. Sue is also an experienced maths teacher and has been a secondary mathematics teacher at all levels in government schools in Victoria.

Wendy Bodey

Wendy Bodey draws on diverse experience gained working at Curriculum Corporation, the Australian Council for Educational Research and as a teacher. Wendy has authored a variety of assessment materials and resources and has played key roles in national and jurisdiction assessment programs including the inaugural National Assessment Program Literacy and Numeracy (NAPLAN) tests.

Sarah Hamper

Sarah Hamper is an experienced mathematics teacher and currently teaches in New South Wales at Abbotsleigh School. Sarah has a Bachelor degree in Pure Mathematics and a Master's degree in Education, studying mathematical modelling, problem solving and using real world applications and Information Communication Learning Technologies (ICLT) to aid effective teaching and learning of mathematics. Sarah has also presented a number of professional papers in this area at state and national level.

Test tips

Year 9 is a pivotal year in your education. Tests provide an important source of information to assist you and your teacher to review your learning goals and to ensure that you are on track to meet your future aspirations. Follow the practical advice provided throughout the rich resources in this book and the accompanying NelsonNet website and you will be prepared to successfully sit the NAPLAN* tests and most tests at the secondary level.

Make mathematics your friend

Make sure you are familiar with the mathematical language and conventions expected at this level. Often words have different meanings in mathematics from their use in everyday life. A simple example of this is the word 'product': you might think of a product as something you make or something that is sold in shops. However, in mathematics, a product is the answer found from multiplying.

Likewise there are various conventions in mathematics, such as the use of notation, that you need to know and apply if you are to arrive at the correct answer to a problem. It is important that you are familiar with the formulae and rules that you may need to use, such as how to calculate area. Your class notes and texts will contain a summary of important information. Use this information to hone your knowledge and skills and build your understanding.

Practise your test strategies

When working through the test, complete the easy questions first then return to any remaining questions. This will allow you more time to think about how to solve the more difficult questions. Remember not to spend too much time on any one question until you have completed the rest of the test. While the test questions will vary from year to year, the range of difficulty and content across the test will be very similar. The practice tests and full-length tests provide a good opportunity for you to get a sense of the type of questions in the test. They will also give you a chance to see what you know and what you might need to revise. You can also use the practice tests as a guide to see how well you are using your time.

Check your answer but not straight away

It's a good idea to complete another question or two before you go back and check your answer to any current question. This could help you avoid repeating an error. Imagine you accidentally calculated 2 x 3 = 5, adding the two numbers rather than multiplying them. This mistake could easily be repeated if you checked your answer straight away. However it is more likely you will notice the mistake if you have cleared your head from your initial calculations when you check it. However, don't wait too long before you check your answer or you may run out of time. A good rule of thumb is to do three questions, then check the first, do another, then check the second and so on throughout the test.

Focus on the entire task

Read the whole question first so that you can focus on the entire task and not get distracted by particular parts of the question. Consider writing the question in your own words or drawing diagrams so that you can clarify the problem that you are about to solve.

Underline or circle important words and numbers in the question to highlight key information. Once you have a clear idea of what the question is asking you to do, you will be in a better position to find the solution. A problem might show a picture of 1500 mL of water in a jug. You are told to add a further 350 mL of water to the jug and then say how much more water is required to have 2 L of water. One common error is to give the amount of water in the jug after the 350 mL is added (1850 mL) rather than subtracting this amount from 2 L to find out how much more water is required (150 mL). Always reread the question to ensure you have a good understanding of the entire task when you are checking your answer.

Check the reasonableness of your answer

Always check your answers for their reasonableness, as this will help you to identify any accidental errors and assist you in eliminating distracters from multiple-choice options. Think about a question where you are required to measure the size of an angle in a triangle. Visually you will be able to recognise whether an angle is acute, obtuse or a right angle. If you measure the angle to be 110° and yet the angle is clearly acute, this will signal your reading is unreasonable. It might simply be that you are reading the inside rather than the outside scale on the protractor. In any case you are alerted to an issue so you can act on it. Likewise, if you were considering the answer options provided in a multiple-choice question, you would quickly be able to eliminate any option that is 90° or larger.

Multiple-choice questions give you the answer

Multiple-choice questions provide the correct answer along with other incorrect options to choose from. Start by doing your own calculations and see if your answer is one of the options provided. If it is, you will feel reasonably confident you have correctly solved the problem. The best way to check this is to make sure that all the other options are incorrect.

Some students think if they are not sure of an answer they should stick with a particular guess, for example always choosing 'B' when unsure. This is not a recommended strategy; it is a myth that 'B' or any other option is usually correct. Most tests have a good mix of correct answers – A, B, C or D, randomly spread across the test with approximately the same number of As, Bs and the other options available. The best way to proceed when you are not sure which option is the correct answer is to eliminate the incorrect options. The following tip provides specific advice on eliminating incorrect options.

Eliminate incorrect options in multiple-choice questions

The options provided in multiple-choice questions are chosen because they are common errors that students make when solving the problem. Therefore, you need to think through the options carefully so you are not distracted by attractive but incorrect options.

First eliminate implausible options. If the question asked you how many hours and minutes there are between 4:49 a.m. and 4:33 p.m. then it would be easy to eliminate any option that was 12 hours or more. The range would need to extend to 4:49 p.m. to be 12 hours; as this is not the case, the number of hours must be equal to 11. This would help you eliminate the following two options: 12 hours and 16 minutes or 12 hours and 44 minutes. The remaining two options are 11 hours and 16 minutes, and 11 hours and 44 minutes. As you have already calculated that the answer is 11 hours and a certain number of minutes, you need to concentrate on the minute component. You may

work this out by adding the number of minutes before the hour (at 4.49 p.m. there are 11 minutes before 5.00 p.m.) to the number of minutes past the hour (at 5.33 p.m. there are 33 minutes past 5.00 p.m.) – a total of 44 minutes. A common error is to simply find the difference between 49 and 33 but this is clearly incorrect. Use strategies such as these to eliminate options and to assist you to confirm the correct answer.

Use the icons

The majority of test questions contained in the two booklets are multiple-choice format while approximately one-quarter of the test questions require you to provide the correct answer. There are icons on the page that tell you what to do, for example 'Shade one bubble' or 'Write your answer in the box'. Follow the icon instructions as you proceed through the test.

There are two separate numeracy tests. You will be able to use a calculator to assist you in one of the test papers so make sure you read the calculator skills section to ensure you can make the best use of your calculator during that part of the test. Throughout these resources distinctive icons will help you identify when you are allowed to use a calculator.

There are icons to alert you to further information available on the Internet. Make the most of this extra information to keep you one step ahead.

And of course, make sure you read all the special tips throughout the resources that are highlighted by the 'Hot tips' icon as these tips relate to things many students forget to do or can do better.

Watch the time but don't hurry

The time available to complete each of the tests is adequate for most students to comfortably complete the test. So be conscious of the time as you work through each test but don't rush your answers.

You will have 40 minutes to complete each test book; this is the equivalent of just over one minute per question. Start by working through the questions you are most confident to answer and do quick checks as you go. Use the remaining test time to tackle questions you are less sure of and to more thoroughly check your answers. It is not wise to spend too much time on any one question until you have completed all other questions in the test.

Be calm and don't panic

On test day make sure you have had a good night's sleep and that you have eaten breakfast so that you are physically prepared to do the test. Be confident in your preparation and you will be ready to tackle the test. Use the pre-test and test day checklists provided to assist your preparation.

It is important that you answer each test question to the best of your ability as this will provide a better guide for your future learning requirements. Tests help to identify your strengths and any potential areas for further development that you may have. It is important that you discuss your results with your teacher to ensure your learning program and goals are the most appropriate for you.

Know the terms, facts and formula

It is important to know mathematical facts such as the names of shapes and solids – do you know the difference between an octagon and a hexagon or the difference between a sphere and a cylinder? Facts such as these are integral to you being able to understand questions posed in the test and to solve problems and apply your reasoning to calculations.

Well before test day revise common mathematical facts and make sure you understand the terms used in maths. Start by familiarising yourself with mathematical terms, for example 'mode' and 'median'. Make sure you know what they mean and where you are likely to use them. Make a point of adding new facts and terms to your current knowledge.

Make sure you have a good knowledge of common formulae such as how to calculate the area of a triangle or the volume of a rectangular prism.

Understand common conventions used in mathematics and how to apply these, such as how to follow the order of operations. For example, in the equation $3 + 4 \times 5 = 23$, the multiplication takes precedence over addition by convention.

Be confident in comparing fractional amounts and being able to work with fractions, such as being able to reduce a fraction to its lowest terms or determine which fraction is the largest when comparing two or more fractions. This includes understanding decimal fractions and how the position of the decimal point indicates relative value.

Knowing basic maths facts and the common formulae and terms used will mean you do not waste time trying to understand what a question is asking you and will be able to maximise the time available to solve the problem.

Prepare well

Some students feel nervous when it comes to test time. The best way to manage this is by ensuring you are well prepared before test day, and then you will have no need for any concern. Find out as much as you can about the tests well before you sit them so that you have an understanding of what is being tested, and how it will be tested. Check your calculator proficiency by reviewing the calculator skills section. Revise important facts, formulae and other information that you will find useful during the test and in building your knowledge.

Work through the three non-calculator and calculator-allowed practice tests provided to give you a sense of what the test will be like and to assist you to identify areas that you may need to revise. Use these sessions to monitor how well you use your time. Refer to the solutions on the NelsonNet website to check your answers. Then complete the four full-length detachable tests (two non-calculator and two calculator-allowed). The full-length tests are of the same length and level of difficulty you can expect to find when you sit the actual tests. The first set is designed for you to use as a practice test, the other set to hand in to your teacher. You will find the solutions to these tests on the NelsonNet website also. Discuss your test results with your teacher and seek their feedback on how well you performed and any areas you might improve on.

13 HOT TIP

Try a different way

Struggling to work an answer out? The best thing to do is circle this question number to flag that you need to come back later and then move on to the next question. Sometimes you will find the problem is easier to solve when you return to it. If you are still struggling to work out the answer try approaching it in a different way. You might draw the problem and attempt to solve it that way. You might use symbols to represent key information in the question and see if that helps. There is usually more than one way to solve a problem so the important thing to remember is to try a different method to work out your answer if your initial method is not working.

Keep up to date with information from your test authority

It is important to keep up with any information about the test. Your test authority will provide regular updates. Contact details for all Australian Test Administration Authorities for the NAPLAN* tests can be found at:

www.naplan.edu.au/test_administration_authorities.html.

Other general information about the tests and specific information such as test dates can be found at:

www.naplan.edu.au.

Pre-test checklist

Use this checklist to ensure you are prepared to successfully sit the tests.

	Activity	Main Resource
☐	Revise important facts, formulae and terms used in numeracy and be familiar with common conventions used	Class texts and notes
☐	Know how to use your calculator	Calculator skills
☐	Build your skills and knowledge informally, too	Solve puzzles and play games Take notice of everyday maths
☐	Complete the six practice tests to get a clear idea about the tests and questions types	Complete the six practice tests
☐	Check your answers against the solutions to evaluate your strengths and any areas you need to revise	Practice test solutions
☐	Get advice on tackling the tests	Test tips
☐	Complete Full-length Tests 7 and 8, the same length and level of difficulty you can expect in the test	Full-length Tests 7 and 8
☐	Check your answers against the solutions to evaluate your strengths and any areas you need to revise	Full-length Tests 7 and 8 solutions
☐	Complete Full-length Tests 9 and 10 and hand in to your teacher	Full-length Tests 9 and 10
☐	Ask your teacher for feedback on your performance and use of time	You and your teacher
☐	Keep updated on test information from your assessment authority	Test Administration Authority

9780170470674

Test day checklist

Use this checklist to ensure you are prepared on test day.

Day before the test

	Activity
☐	Calculator, HB or 2B pencils, an eraser and a sharpener ready to take
☐	Have a good night's sleep

Test morning

	Activity
☐	Have breakfast
☐	Take watch, calculator, pencils, eraser and sharpener
☐	Arrive at school, or the testing venue, well before the session commences

During each test

	Activity
☐	Be confident in your preparation
☐	Monitor your time during the test
☐	Work through the test, completing easy questions first
☐	Read each question carefully, underline important words
☐	Don't spend too much time on any one question
☐	Circle the question number of any question you need to return to
☐	Make sure your written answers are legible
☐	Only write in the box or on the lines provided
☐	Select the correct option in a multiple-choice question, check the other options are incorrect
☐	Choose one option only and colour in the bubble or box completely
☐	If you change your answer, rub out the other answer completely
☐	Go back to complete unanswered questions
☐	Wait a question or two before going back to check your answer
☐	Check your work, make sure you haven't skipped any questions

After the test

	Activity
☐	Discuss your results with your teacher
☐	Identify your strengths and any potential areas to revise
☐	Consider these results together with other evidence of your progress
☐	Review learning goals to ensure they are appropriate

Calculator skills

NOTE: The following advice is based on common scientific calculators in use throughout Australian schools.

Basic keys

Key	Use	Key	Use
$+$, $-$, $\times$, $\div$	Basic operations	(–) or +/–	Enters negative numbers
$=$	Equals sign, gives the answer	▭ or $a^b/_c$	Enters fractions
·	Decimal point	( )	Enters parentheses
DEL	Deletes previous entry	x^2	Squares a number
ANS	Retrieves previous answer	$\sqrt{\ }$	Finds the square root of a number
↕ ↔	Allows us to move around the screen	x^3	Cubes a number
MODE or SHIFT or 2ndF	Access other operations	$^3\sqrt{\ }$	Finds the cube root of a number

Basic operations and using brackets

Question	Calculator steps	Answer
246 + 107 =	246 $+$ 107 $=$	353
462 – 289 =	462 $-$ 289 $=$	173
6 × 14 =	6 $\times$ 14 $=$	84
1463 ÷ 7 =	1463 $\div$ 7 $=$	209
8 × (15 + 12) =	8 $\times$ (15 $+$ 12) $=$	216
[(72–6) ÷ (25+8)] × 24 =	[(72 $-$ 6) $\div$ (25 $+$ 8)] $\times$ 24 $=$	48
Find the average of 7.1, 3.6, 8 and 4.5	(7.1 $+$ 3.6 $+$ 8 $+$ 4.5) $\div$ 4 $=$	5.8

Integers

Question	Calculator steps	Answer
-6 + 15 =	+/– 6 $+$ 15 $=$	9
-16 × 12 =	+/– 16 $\times$ 12 $=$	-192
8 – (-12) =	8 $-$ +/– 12 $=$	20
-11 – 7 × -3 =	+/– 11 $-$ 7 $\times$ +/– 3 $=$	10

9780170470674

Decimals

Question	Calculator steps	Answer
17.622 + 5.4 – 8.39 =	17 [·] 622 [+] 5 [·] 4 [–] 8 [·] 39 [=]	14.632
23.1 × 0.82 =	23 [·] 1 [x] 0 [·] 82 [=]	189.42
1.72 ÷ 0.8 =	1 [·] 72 [÷] 0 [·] 8 [=]	0.92
\$7.60 × 9 =	7 [·] 6 [x] 9 [=]	\$68.40
0.9 of 460 m =	0 [·] 9 [x] 460 [=]	414 m
55% × \$186 =	0 [·] 55 [x] 186 [=]	\$102.30
216% of 35 =	2 [·] 16 [x] 35 [=]	75.6

Powers and roots

Question	Calculator steps	Answer
$15^2 =$	15 [x^2] [=] or 15 [x] 15 [=]	225
$9^3 =$	9 [x^3] [=] or 9 [x] 9 [x] 9 [=]	729
$\sqrt{289} - \sqrt[3]{343} =$	[$\sqrt{\ }$] 289 [–] [$\sqrt[3]{\ }$] 343 [=]	10

Operations with fractions

Question	Calculator steps	Answer
Simplify: $\frac{56}{84}$	56 [$\frac{\square}{\square}$] 84 [=]	$\frac{2}{3}$
Convert $2\frac{3}{8}$ to an improper fraction	2 [$\frac{\square}{\square}$] 3 [$\frac{\square}{\square}$] 8 [=] [SHIFT] [$\frac{\square}{\square}$]	$\frac{19}{8}$
Convert $\frac{11}{5}$ to a mixed numeral	11 [$\frac{\square}{\square}$] 5 [=]	$2\frac{1}{5}$
$3\frac{2}{5} + 4\frac{3}{10} =$	3 [$\frac{\square}{\square}$] 2 [$\frac{\square}{\square}$] 5+4 [$\frac{\square}{\square}$] 3 [$\frac{\square}{\square}$] 10 [=]	$7\frac{1}{10}$
$\frac{2}{5}$ *of* 250m =	2 [$\frac{\square}{\square}$] 5 [x] 250 [=]	100 m

Test 1: Non-calculator

Instructions

- A correct answer scores 1 mark, and an incorrect answer scores 0.
- Marks are not deducted for incorrect answers.
- No marks are given if more than one answer alternative is shaded.
- Choose the alternative which most correctly answers the question and shade in the box next to it.

QUESTION 1

SHADE ONE BOX

What is the number fifty-five thousand and twenty?

☐ 5520   ☐ 550 020   ☐ 55 020   ☐ 5 500 020

QUESTION 2

SHADE ONE BOX

The column graph below shows information about children's use of the Internet.

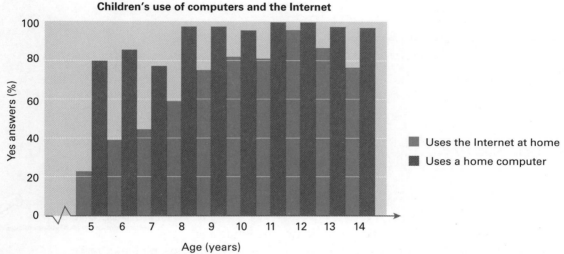

Which of the following statements is correct?

☐ More than 70% of 8 year olds use the Internet at home.

☐ Almost all children over 8 years have a home computer.

☐ More than 80% of 14 year olds use the Internet at home.

☐ More than 90% of 7 year olds have a home computer.

QUESTION 3

 WRITE YOUR OWN ANSWER

Write 15.75 cm as metres.

☐

QUESTION 4

SHADE ONE BOX

What can 2p + 3q + 4p + 6q – p be written as?

☐ 5p + 9q   ☐ 14pq   ☐ 13pq   ☐ 6p + 9q

QUESTION 5

SHADE ONE BOX

Which letter on the number line represents -4?

D ☐   C ☐   A ☐   B ☐

0   2

QUESTION 6

SHADE ONE BOX

What is the best name for this shape?

☐ rhombus

☐ trapezium

☐ kite

☐ parallelogram

QUESTION 7

SHADE ONE BOX

Jacob has several socks in his drawer. Five are black, two are white and three are blue. What is the chance of Jacob pulling out a white sock if he doesn't look first?

☐ $\frac{1}{2}$ ☐ $\frac{1}{5}$ ☐ $\frac{3}{10}$ ☐ $\frac{1}{3}$

QUESTION 8

SHADE ONE BOX

What is the best estimate of 34 × 56?

☐ 1200   ☐ 1500   ☐ 1700   ☐ 1800

9780170470674

QUESTION 9

WRITE YOUR OWN ANSWER

The area of this rectangle is 63 cm^2.

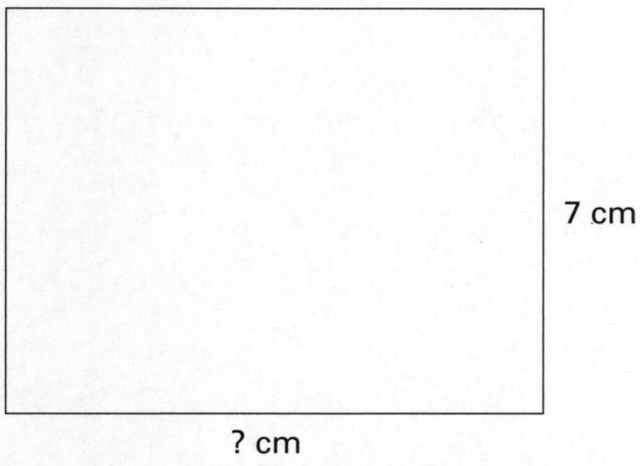

7 cm

? cm

What is the length of the rectangle?

[ ] cm

QUESTION 10

WRITE YOUR OWN ANSWER

Lotte cuts her birthday cake into 12 equal slices. She eats one-quarter of the cake. How many slices are left?

[ ] slices

QUESTION 11

SHADE ONE BOX

What is the value of x in the diagram?

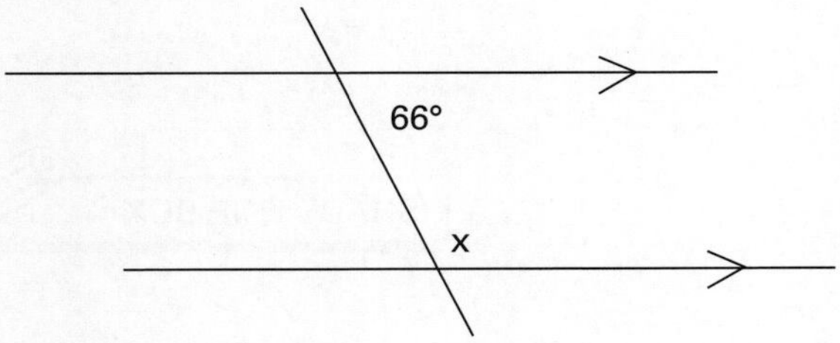

66°

x

[ ] 124° [ ] 66° [ ] 24° [ ] 114°

QUESTION 12

SHADE ONE BOX

A chemical alloy is made up of 12 per cent nickel, 5 per cent gold, 16 per cent copper and the remainder is iron. What percentage of iron is in the alloy?

[ ] 87%

[ ] 77%

[ ] 67%

[ ] 57%

QUESTION 13

SHADE ONE BOX

$2^3 + 4^2 - 0^4 = ?$

[ ] 14 [ ] 24 [ ] 10 [ ] 22

QUESTION 14

 SHADE ONE BOX

I multiply a number by 3, square the result, add 5 and then multiply by 4. The answer is 15.

Which algebraic expression represents this process?

☐ $3a^2 + 5 \times 4 = 15$

☐ $9a^2 + 20 = 15$

☐ $4(9a^2 + 5) = 15$

☐ $4(3a^2 + 5) = 15$

QUESTION 15

 WRITE YOUR OWN ANSWER

What is the gradient (m) of this line?

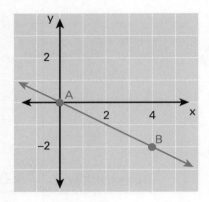

m =

QUESTION 16

 SHADE ONE BOX

A builder is looking at a model of a building site.

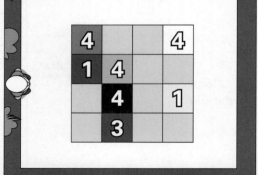

Which of the following represents what he sees?

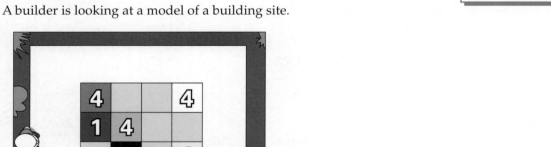

☐ 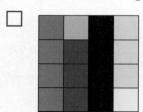   ☐    ☐ 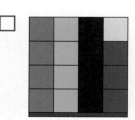   ☐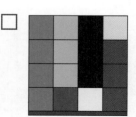

QUESTION 17

Which of the following data collected can be classified as continuous data?

☐ teams in the final series

☐ favourite pets

☐ daily maximum temperatures in September

☐ lengths of dogs' tails

QUESTION 18

This dot plot shows the ages of 40 people enrolled for a judo class.

Age (years)

What is the median age of the people enrolled?

☐ 18   ☐ 25   ☐ 27   ☐ 28

QUESTION 19

What percentage of the diagram is **not** shaded?

☐ 24%   ☐ 36%   ☐ 40%   ☐ 60%

SHADE ONE BOX

QUESTION 20

What is the area of this triangle?

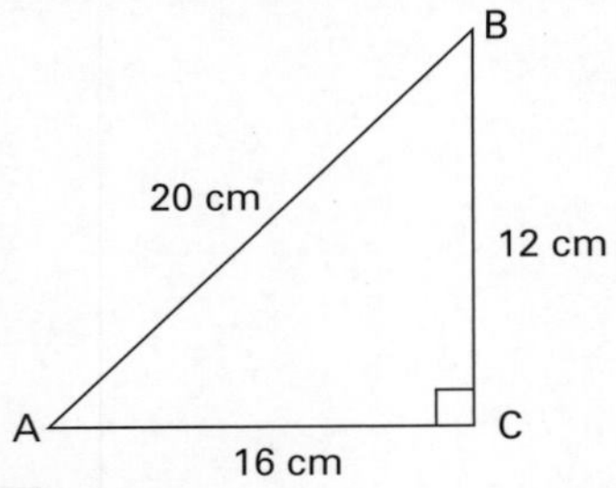

- [ ] 240 cm^2
- [ ] 192 cm^2
- [ ] 120 cm^2
- [ ] 96 cm^2

WRITE YOUR OWN ANSWER

QUESTION 21

Emma, who lives in Ballarat, Victoria, wants to call her friend George who lives in Darwin. Victoria is 1.5 hours ahead of the Northern Territory.
If Emma calls at 11.00 a.m. Australian Eastern Daylight-saving Time, what time will it be for George?
Give your answer in 24-hour time format.

[ ]

SHADE ONE BOX

QUESTION 22

Zahra made the following measurements when trying to work out the height of a building.

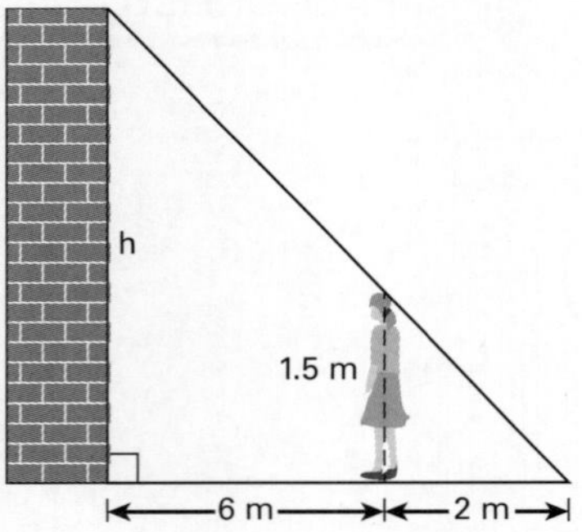

What is the height of the building?

- [ ] 6 m
- [ ] 5 m
- [ ] 4.5 m
- [ ] 3.5 m

SHADE ONE BOX

QUESTION 23

Which fraction has the same value as $3\frac{1}{3}$?

- [ ] $\frac{10}{3}$
- [ ] $\frac{4}{3}$
- [ ] 3.3
- [ ] 3.1

9780170470674

QUESTION 24

Which of the following best describes the translation of the grey figure to the blue figure in the diagram below?

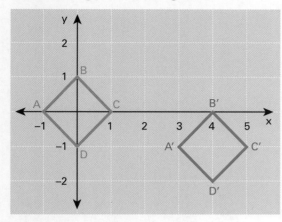

- [ ] 4 horizontally and -2 vertically
- [ ] 1 horizontally and 4 vertically
- [ ] -4 horizontally and 2 vertically
- [ ] 4 horizontally and -1 vertically

QUESTION 25

What is the ratio of black to blue circles in the diagram below?

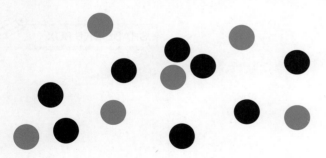

- [ ] 4:3
- [ ] 4:7
- [ ] 3:4
- [ ] 3:7

QUESTION 26

If $\frac{3x}{4} - 1 = 8$, what is the value of x?

QUESTION 27

Two normal dice are tossed. What is the probability of getting a 10 as the sum of the numbers on the two dice?

- [ ] $\frac{1}{12}$
- [ ] $\frac{5}{18}$
- [ ] $\frac{5}{36}$
- [ ] $\frac{1}{6}$

QUESTION 28

SHADE ONE BOX

In the two-way table shown below, which represents a class of students,
G represents girls and L represents speaking a language other than English at home.

	L	L'	
G	5	9	14
G'	3	10	13
	8	19	27

Note: G' means not a girl.

What is the number of girls who speak English at home?

☐ 14   ☐ 5   ☐ 9   ☐ 19

QUESTION 29

WRITE YOUR OWN ANSWER

What is the volume of this prism?   cm³

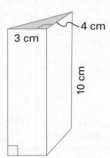

4 cm

3 cm

10 cm

QUESTION 30

SHADE ONE BOX

The table of values for $y = 6 - 2x^2$ is:

☐
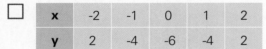

x	-2	-1	0	1	2
y	10	8	6	4	2

☐
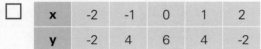

x	-2	-1	0	1	2
y	2	-4	-6	-4	2

☐
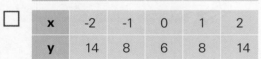

x	-2	-1	0	1	2
y	-2	4	6	4	-2

☐

x	-2	-1	0	1	2
y	14	8	6	8	14

QUESTION 31

SHADE ONE BOX

There are 5400 mice in a paddock with an area of 90 hectares (ha). What is
the population density?

☐ 5400 mice/ha

☐ 60 mice/ha

☐ 600 mice/ha

☐ 90 mice/ha

9780170470674

QUESTION 32

Five computers are to be connected together with the minimum length of fibre optic cable. The possible links are shown in the diagram.

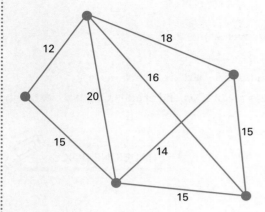

What is the minimum length of cable required?

☐ 56 m   ☐ 60 m   ☐ 58 m   ☐ 57 m

Test 2: Calculator allowed

Instructions

- A correct answer scores 1 mark, and an incorrect answer scores 0.
- Marks are not deducted for incorrect answers.
- No marks are given if more than one answer alternative is shaded.
- Choose the alternative which most correctly answers the question and shade in the box next to it.

QUESTION 1

SHADE ONE BOX

Which time is the same as the time shown on this digital clock?

19:20

- [ ] 7:20 a.m.
- [ ] 9:20 a.m.
- [ ] 7:20 p.m.
- [ ] 9:20 p.m.

QUESTION 2

WRITE YOUR OWN ANSWER

What is the missing number?

$8 \times 57 = 19 \times ?$ ☐

QUESTION 3

SHADE ONE BOX

What is the value of m° in this triangle?

m°

- [ ] 30°
- [ ] 45°
- [ ] 60°
- [ ] 120°

QUESTION 4

SHADE ONE BOX

Which one of these is equivalent to $\frac{4}{5}$?

- [ ] 80%
- [ ] $\frac{18}{20}$
- [ ] 0.75
- [ ] $\frac{44}{45}$

9780170470674

QUESTION 5

WRITE YOUR OWN ANSWER

For six games in a basketball season, David has scored the following points: 6, 11, 16, 8, 12 and 7.

What is the average number of points David scored per game?

QUESTION 6

WRITE YOUR OWN ANSWER

2568 − 719 = ?

QUESTION 7

WRITE YOUR OWN ANSWER

Nishan ran a distance of 100 m in 16 seconds. What is Nishan's average speed (in metres per second)?

metres per second

QUESTION 8

SHADE ONE BOX

A tin contains 10 yellow, 3 red and 7 white jellybeans. Without looking, Kai takes one jellybean from the tin. What is the probability that Kai chooses a yellow jellybean?

☐ 0.1   ☐ 0.2   ☐ 0.5   ☐ 0.7

QUESTION 9

SHADE ONE BOX

Anousha and Ethan share $175 in a ratio of 4:3. How much does Ethan receive?

☐ $100   ☐ $75   ☐ $50   ☐ $25

QUESTION 10

SHADE ONE BOX

Darcy uses this rule to work out the next number in a pattern: multiply by 9 and then add 4.

The first three terms of this pattern are: 13, 22, 31, …

What is the seventh term in this pattern?

☐ 63   ☐ 66   ☐ 67   ☐ 71

QUESTION 11

SHADE ONE BOX

Which fraction has the same value as $7\frac{1}{2}$?

☐ $\frac{21}{2}$

☐ $\frac{10}{2}$

☐ $\frac{14}{2}$

☐ $\frac{15}{2}$

QUESTION 12

What is the size of angle x?

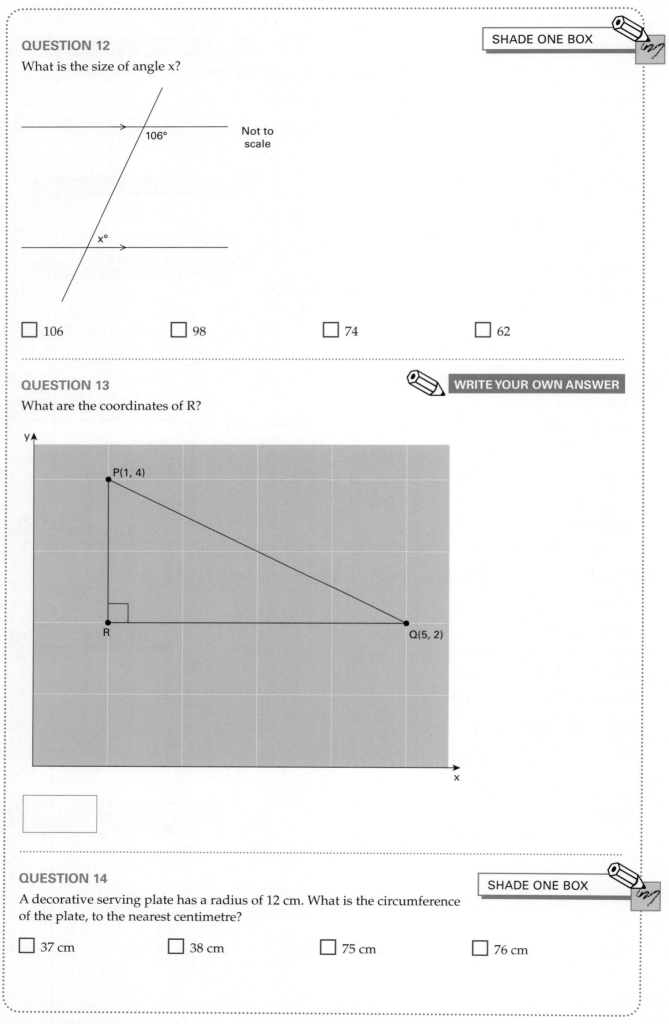

Not to scale

106°

x°

☐ 106   ☐ 98   ☐ 74   ☐ 62

QUESTION 13

What are the coordinates of R?

P(1, 4)

R

Q(5, 2)

QUESTION 14

A decorative serving plate has a radius of 12 cm. What is the circumference of the plate, to the nearest centimetre?

☐ 37 cm   ☐ 38 cm   ☐ 75 cm   ☐ 76 cm

QUESTION 15

The map below shows the layout of a showground.

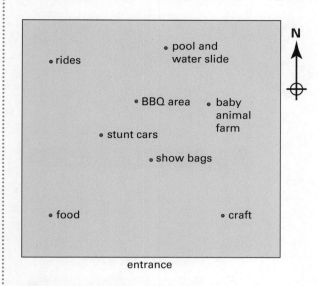

entrance

If Will is standing at the show bags, in what direction are the rides?

- [ ] south-east
- [ ] south-west
- [ ] north-east
- [ ] north-west

QUESTION 16

What is the square root of 500?

- [ ] between 15 and 20
- [ ] between 20 and 25
- [ ] between 60 and 90
- [ ] between 200 and 250

QUESTION 17

The graph shows the value of two cars and their age in years. Which one of these statements is true?

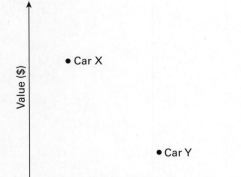

- [ ] Car X is an older car and less expensive than car Y.
- [ ] Car Y is a newer car and less expensive than car X.
- [ ] Car Y is an older car and worth more than car X.
- [ ] Car X is a newer car and worth more than car Y.

QUESTION 18

Which diagram below shows the net of a square pyramid?

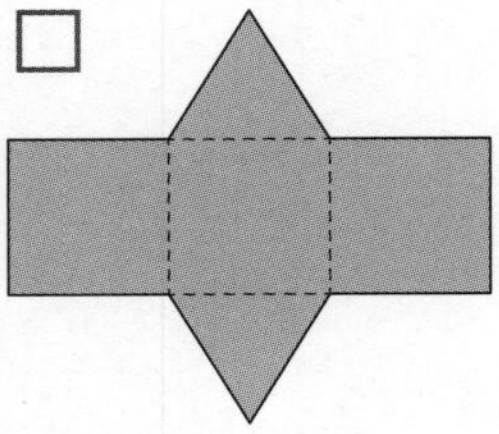

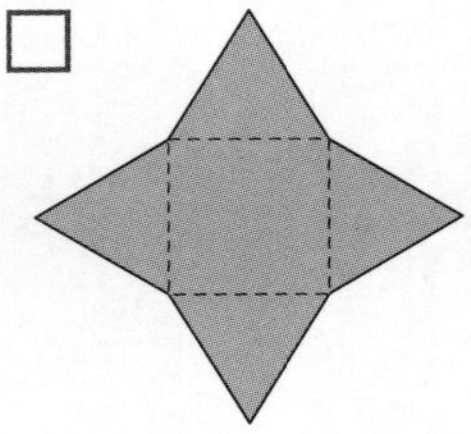

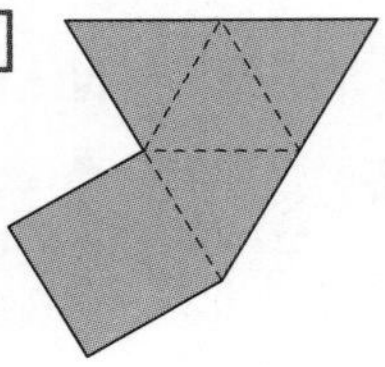

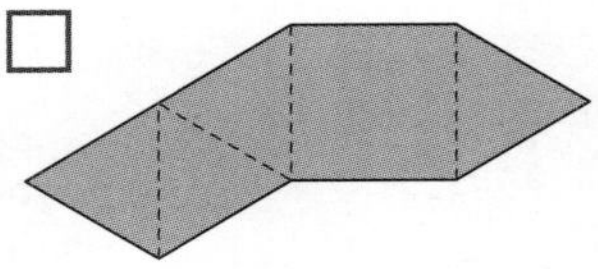

SHADE ONE BOX

☐   ☐   ☐   ☐

QUESTION 19

WRITE YOUR OWN ANSWER

A triangle has an area of 357 cm^2 and a perpendicular height of 21 cm. What is the length of its base?

cm

QUESTION 20

WRITE YOUR OWN ANSWER

If x = 7, what is the value of the expression 2x^2

QUESTION 21

SHADE ONE BOX

A rainwater tank holds 15.5 kL when full. How many litres of rainwater is this?

☐ 155   ☐ 1505   ☐ 1550   ☐ 15 500

QUESTION 22

SHADE ONE BOX

Nick bought a new lounge at a furniture sale. He was given a 15 per cent discount off the original price. The original price of the lounge was $1600. How much did Nick pay for the lounge?

☐ $240   ☐ $320   ☐ $1360   ☐ $1440

QUESTION 23

SHADE ONE BOX

Which one of these gives the largest value?

☐ $\sqrt{1.44}$

☐ 0.27 ÷ 0.3

☐ 4.6 − 3.52

☐ 0.73 + 0.5 + 0.14

QUESTION 24

6x + 4y − x = ?

SHADE ONE BOX

☐ 5x + 4y   ☐ 9xy   ☐ 6 + 4y   ☐ 10xy − x

QUESTION 25

SHADE ONE BOX

Tyneville is 6.2 cm from Reed Hills on a map.

On the map, 1 cm represents 4 km.

What is the actual distance from Tyneville to Reed Hills?

☐ 24.8 km   ☐ 15.5 km   ☐ 2.2 km   ☐ 0.65 km

QUESTION 26

WRITE YOUR OWN ANSWER

A regular pentagon has an angle sum of 540°. What is the size of the angle a°?

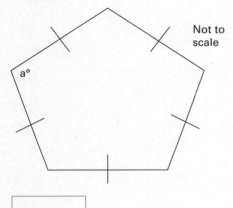

Not to scale

a°

☐

QUESTION 27

SHADE ONE BOX

Which diagram shows a rhombus?

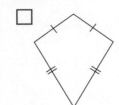 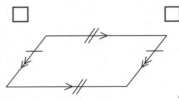 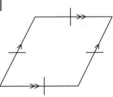

☐   ☐   ☐   ☐

QUESTION 28

SHADE ONE BOX

Hannah randomly takes one ball from one of the bags shown below.
Which bag would Hannah need to use to have a one in three chance of drawing out a white ball?

☐   ☐   ☐   ☐

   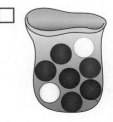

QUESTION 29

Students' total scores for correct answers in a spelling quiz of ten questions are shown below:

7, 5, 6, 4, 2, 7, 6, 2, 3, 7, 7, 6.

What is the modal score?

QUESTION 30

 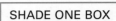

Last month it rained on 6 days and was sunny on 24 days. For what percentage of the month did it rain?

 80%   75%   25%   20%

QUESTION 31

A rectangular prism is shown below.

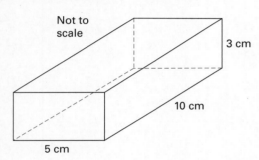

What is the surface area of this rectangular prism?

QUESTION 32

PQRS is a rectangle. If PQ = 2 × QR, what is the value of a?

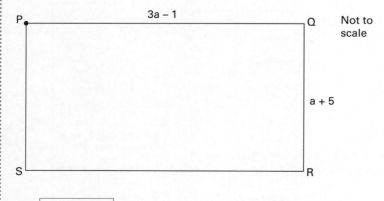

a =

Test 3: Non-calculator

Instructions

- A correct answer scores 1 mark, and an incorrect answer scores 0.
- Marks are not deducted for incorrect answers.
- No marks are given if more than one answer alternative is shaded.
- Choose the alternative which most correctly answers the question and shade in the box next to it.

QUESTION 1

SHADE ONE BOX

$3x \times 4x = ?$

☐ $7x$ ☐ $12x$ ☐ $7x^2$ ☐ $12x^2$

QUESTION 2

WRITE YOUR OWN ANSWER

The arrow points to a position on the number line. What number is represented by A?

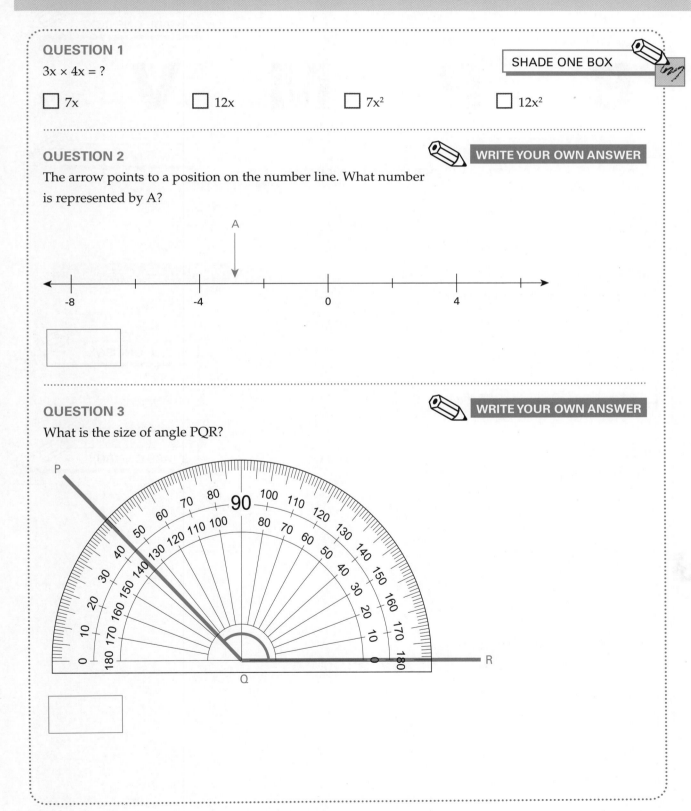

QUESTION 3

WRITE YOUR OWN ANSWER

What is the size of angle PQR?

QUESTION 4

SHADE ONE BOX

How many hours and minutes are between 11.25 a.m. and 1.14 p.m. on the same day?

- [ ] 1 hour 14 minutes
- [ ] 1 hour 49 minutes
- [ ] 2 hour 14 minutes
- [ ] 2 hour 49 minutes

QUESTION 5

SHADE ONE BOX

Which letter does not have a line of symmetry?

- [ ] **B**
- [ ] **P**
- [ ] **M**
- [ ] **V**

QUESTION 6

SHADE ONE BOX

Henry has 18 orange flowers and 12 pink flowers. What fraction of the flowers is pink?

- [ ] $\frac{2}{3}$
- [ ] $\frac{2}{5}$
- [ ] $\frac{3}{8}$
- [ ] $\frac{3}{10}$

QUESTION 7

WRITE YOUR OWN ANSWER

How many vertices does a cube have? ☐ vertices

QUESTION 8

SHADE ONE BOX

$1.4 \times 0.6 = ?$

- [ ] 84
- [ ] 8.4
- [ ] 0.84
- [ ] 0.084

QUESTION 9

SHADE ONE BOX

What type of angles are these?

- [ ] cointerior
- [ ] alternate
- [ ] scalene
- [ ] corresponding

QUESTION 10

This object was made using identical cubes.

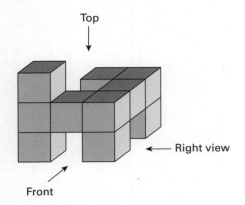

Which drawing shows the view from the right side?

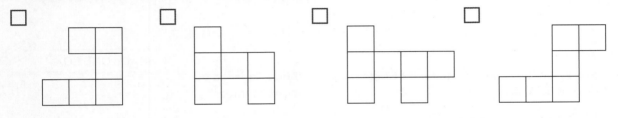

QUESTION 11

 WRITE YOUR OWN ANSWER

This graph shows which students participate in swimming and tennis in Year 9.

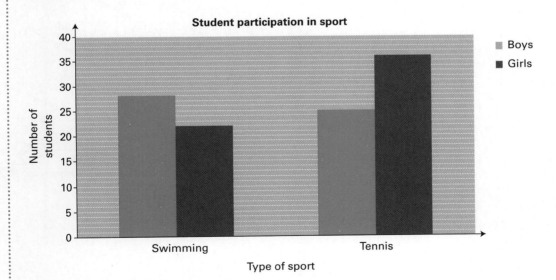

Use the information in the graph to complete this table.

	Swimming	Tennis
Boys	28	25
Girls		36

QUESTION 12

SHADE ONE BOX

Which one of the following represents the equation of the graph below?

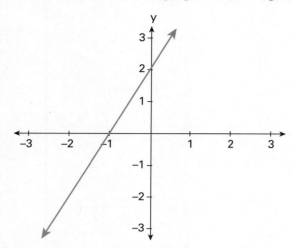

☐ y = 2x + 2   ☐ y = 2x − 1   ☐ y = x + 2   ☐ y = x − 1

QUESTION 13

SHADE ONE BOX

A spinner is in the shape of a regular hexagon. Each section is labelled black, yellow or red. Which spinner below shows the following probabilities?

Probability of landing on black = $\frac{1}{2}$

Probability of landing on yellow = $\frac{1}{3}$

Probability of landing on red = $\frac{1}{6}$

☐   ☐   ☐   ☐

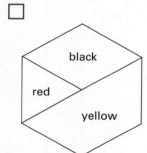 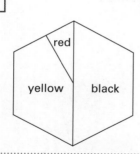 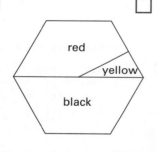 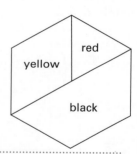

QUESTION 14

SHADE ONE BOX

Look at the object shown below

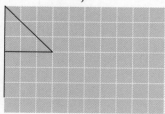

Which diagram below shows the same object multiplied by a scale factor of $\frac{1}{3}$?

☐   ☐   ☐   ☐

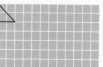 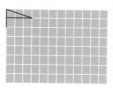 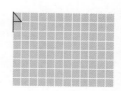

9780170470674

QUESTION 15

This jug contains some orange juice.

If an extra 800 mL of orange juice is added to the jug,
how many litres of orange juice will then be in the jug?

	litres

QUESTION 16

SHADE ONE BOX

$\frac{5}{6} \times \frac{3}{4} = ?$

☐ $\frac{4}{5}$ ☐ $\frac{3}{4}$ ☐ $\frac{2}{3}$ ☐ $\frac{5}{8}$

QUESTION 17

SHADE ONE BOX

Ki Min drew this shape.

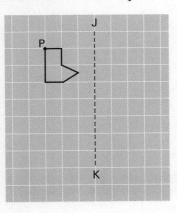

He reflects it in the line JK and then rotates it 90° in an anticlockwise direction about point P.

What would it look like after it has been reflected and rotated?

☐ 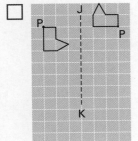   ☐ 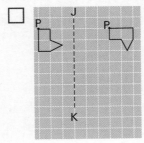   ☐ 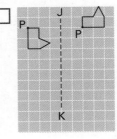   ☐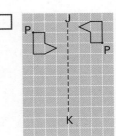

What is the volume of this solid?

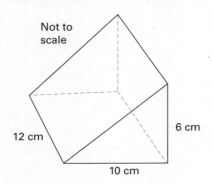

Not to scale

12 cm

6 cm

10 cm

cm³

A number is multiplied by 7 and then 5 is added. The answer is 61.
What is the number?

Mawson Antarctic Base – temperatures in June 2009

	Minimum temperature (°C)
Lowest temperature	-25.6
Highest temperature	-10.1

What was the difference between the lowest and highest minimum temperature in June?

Alexander created a table to show the favourite colour of students in Year 9.

Favourite colour	Number of students
Red	12
Blue	29
Purple	51
Green	?
Pink	28
Total	150

What percentage of Year 9 students like green?

QUESTION 22

SHADE ONE BOX

Over a two-week period, Amber planted x trees in a bush regeneration area. In the same time period, Tyus planted 25 more trees than Amber. Together they planted a total of 116 trees. Which equation represents this information?

☐ x − 25 = 116

☐ x + 25 = 116

☐ 2x + 25 = 116

☐ 2(x − 25) = 116

QUESTION 23

SHADE ONE BOX

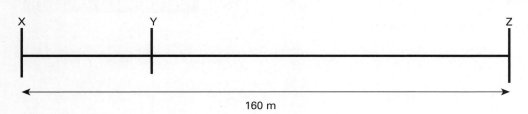

160 m

The distance from Y to Z is three times the distance from X to Y. The distance from X to Z is 160 m. What is the distance from X to Y?

☐ 24 m   ☐ 30 m   ☐ 40 m   ☐ 45 m

QUESTION 24

SHADE ONE BOX

Kishor drove his car 280 km in 3.5 hours. What was his average speed per hour?

☐ 80 km/h   ☐ 90 km/h   ☐ 75 km/h   ☐ 60 km/h

QUESTION 25

SHADE ONE BOX

This table shows the number of loaves of different types of bread sold in a bakery on one day.

Type of bread	White	Wholemeal	Multigrain	Sourdough	Raisin bread
Number of loaves	170	240	65	95	80

What is the ratio of wholemeal loaves to raisin bread loaves sold on this day?

☐ 3:1

☐ 4:1

☐ 1:3

☐ 1:4

QUESTION 26

Which of the following shows the correct expression for the perimeter of this rectangle?

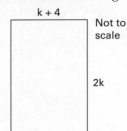

$k + 4$

Not to scale

$2k$

☐ $6k + 8$ ☐ $3k + 4$ ☐ $2k^2 + 4$ ☐ $2k^2 + 8k$

QUESTION 27

How many triangles are there in this diagram?

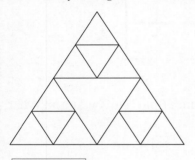

☐ triangles

QUESTION 28

A box contains 30 loaves of bread, each of the same weight.

The box of loaves weighs 6 kg.

Gemma adds 10 more of these loaves to the box.

The weight of the 40 loaves now in the box is ☐ kg.

QUESTION 29

The number of computers Sophie has sold each week for the last 14 weeks is shown in the stem-and-leaf plot below.

Stem	Leaf
6	0 1 4
7	2 2 5 9
8	1 3 3 3
9	7 8 8

Key

6|0 = 60 computers

What is the range of the data in this stem-and-leaf plot?

☐ 60   ☐ 83   ☐ 38   ☐ 26

QUESTION 30

Write 120 as a product of its prime factors.

QUESTION 31

One Australian dollar buys 0.6 Euros. How many Australian dollars would be equivalent to 90 Euros, using this exchange rate?

☐ $175 ☐ $150 ☐ $54 ☐ $60

QUESTION 32

This is a map of a camping and recreation area.

If 1 unit is equivalent to 50 m, what is the shortest actual distance between the canoeing meeting point and campsite B?

☐ m

Test 4: Calculator allowed

Instructions

- A correct answer scores 1 mark, and an incorrect answer scores 0.
- Marks are not deducted for incorrect answers.
- No marks are given if more than one answer alternative is shaded.
- Choose the alternative which most correctly answers the question and shade in the box next to it.

QUESTION 1

SHADE ONE BOX

Which of the following sets of numbers is arranged from largest to smallest?

- ☐ 1.023, 10.23, 102.3, 1032
- ☐ 1.023, 102.3, 1032, 10.23
- ☐ 1032, 102.3, 10.23, 1.023
- ☐ 1032, 10.32, 102.3, 1.02

QUESTION 2

WRITE YOUR OWN ANSWER

John measured the length of a table as 1.20 m.
What is this measurement in centimetres?

☐ cm

QUESTION 3

SHADE ONE BOX

Which of the following numbers is a prime number?

☐ 6   ☐ 7   ☐ 8   ☐ 9

QUESTION 4

SHADE ONE BOX

A quadrilateral that has two pairs of parallel sides is best called

- ☐ a rectangle
- ☐ a rhombus
- ☐ a square
- ☐ a parallelogram

QUESTION 5

In 2008, one of the questions asked in the Census-at-School survey was 'What are your favourite types of music?' The following data is from a random sample of 20 students.

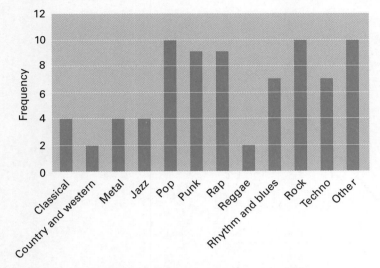

Which of the following statements is true about the data?

☐ Fewer students said their favourite was classical than reggae.

☐ Fewer students said their favourite was pop than rock.

☐ Fewer students said their favourite was rhythm and blues than punk.

☐ Fewer students said their favourite was techno than metal.

QUESTION 6

 WRITE YOUR OWN ANSWER

Mandip bought a packet of 15 cards for $26.25.
What is the average cost of a card?

QUESTION 7

SHADE ONE BOX

The ratio of Max's age to Charlotte's is 3:5. Max is six years old. How old is Charlotte?

☐ 12 years ☐ 10 years ☐ 6 years ☐ 4 years

QUESTION 8

 WRITE YOUR OWN ANSWER

The area of a lake that is covered by algae doubles every day.
The graph below shows the area covered each day.

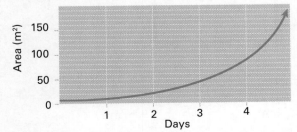

Use the graph to find the number of days it takes to cover 100 m². Give your answer to the nearest day.

Approximately [ ] days

QUESTION 9

What is the size of angle ABC?

WRITE YOUR OWN ANSWER

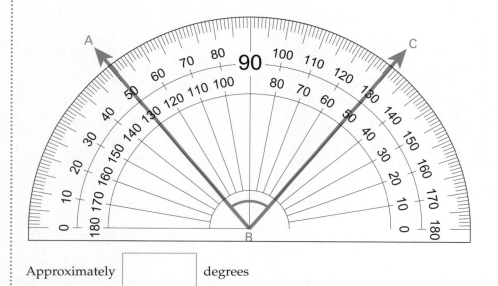

Approximately [ ] degrees

QUESTION 10

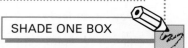
SHADE ONE BOX

Jamil has decided that she will have 'Bed 4' as hers when she and her family move into their new house.

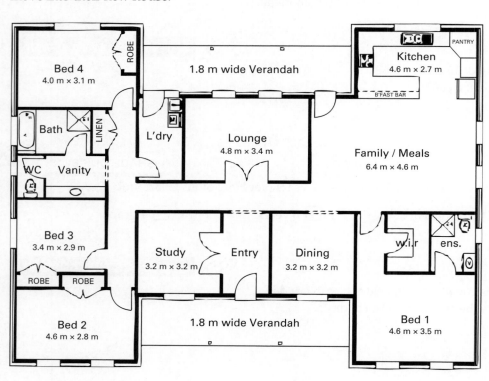

What is the floor area of Jamil's bedroom including the robe?

☐ 14.2 m²   ☐ 7.1 m²   ☐ 12.4 m²   ☐ 9.86 m²

QUESTION 11

WRITE YOUR OWN ANSWER

The arrow points to a position on the number line.

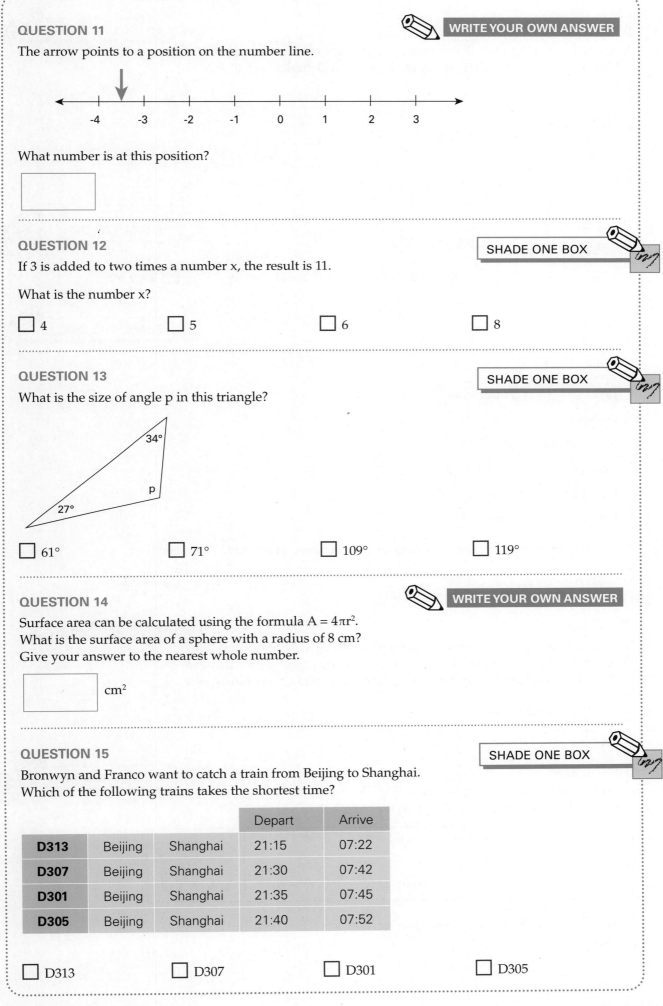

What number is at this position?

QUESTION 12

SHADE ONE BOX

If 3 is added to two times a number x, the result is 11.

What is the number x?

☐ 4   ☐ 5   ☐ 6   ☐ 8

QUESTION 13

SHADE ONE BOX

What is the size of angle p in this triangle?

☐ 61°   ☐ 71°   ☐ 109°   ☐ 119°

QUESTION 14

WRITE YOUR OWN ANSWER

Surface area can be calculated using the formula $A = 4\pi r^2$.
What is the surface area of a sphere with a radius of 8 cm?
Give your answer to the nearest whole number.

☐ cm²

QUESTION 15

SHADE ONE BOX

Bronwyn and Franco want to catch a train from Beijing to Shanghai.
Which of the following trains takes the shortest time?

			Depart	Arrive
D313	Beijing	Shanghai	21:15	07:22
D307	Beijing	Shanghai	21:30	07:42
D301	Beijing	Shanghai	21:35	07:45
D305	Beijing	Shanghai	21:40	07:52

☐ D313   ☐ D307   ☐ D301   ☐ D305

QUESTION 16

Megastore is having a sale. Everything is 15 per cent off the marked price.
If Phoebe pays cash, how much will she pay for a CD marked at $29.95?

☐ $34.45

☐ $28.45

☐ $25.45

☐ $29.80

QUESTION 17

Which one of these is equal to 2^3?

☐ 2×3 ☐ $2 \times 2 \times 2$ ☐ 3×3 ☐ $2 + 2 + 2$

QUESTION 18

The table below shows the population in the five most populous cities in
the United States of America in 2008.

City	Population
New York	8 363 710
Los Angeles	3 833 995
Chicago	2 853 114
Houston	2 242 193
Phoenix	1 567 924

How many more people live in Los Angeles than Houston and Phoenix combined?

☐ 23 878   ☐ 23 882   ☐ 24 878   ☐ 24 882

QUESTION 19

Mannix sold teddy bears to raise money for a children's charity.
Each teddy bear was sold for $5. The charity received 60c for each teddy bear
sold. If Mannix raised $90 for the charity, how many teddy bears did he sell?

☐ teddy bears

QUESTION 20

Jason is an athlete in training for a triathlon. He divides his
weekly training sessions up as shown in the bar graph below.

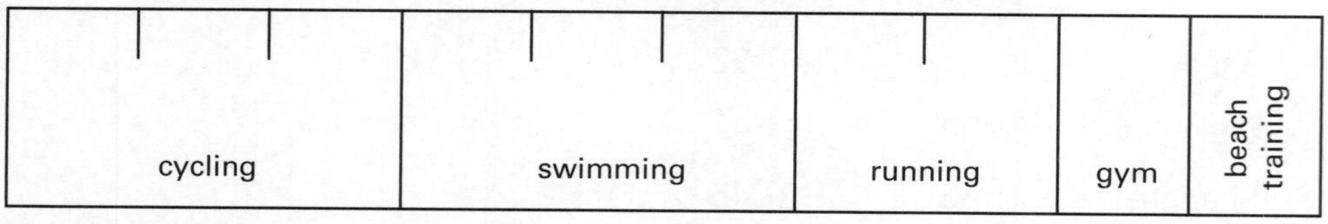

cycling	swimming	running	gym	beach training

If Jason spends 18 hours each week running, how long does he spend cycling?

☐ 12 hours   ☐ 15 hours   ☐ 27 hours   ☐ 30 hours

QUESTION 21

SHADE ONE BOX

A good darts player attempts to hit the bullseye. From 20 attempts she manages to hit the bullseye a total of 7 times. If she makes another 55 attempts, how many bullseyes could she expect?

☐ 19 ☐ 20 ☐ 42 ☐ 12

QUESTION 22

SHADE ONE BOX

Blocks are used to make these L-shaped patterns.

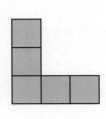 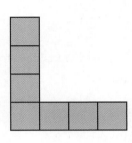

Arm length 1   Arm length 2   Arm length 3

What is the rule connecting the number of tiles t and the arm length a?

☐ $a = t + 1$

☐ $t = a + 1$

☐ $a = 2t + 1$

☐ $t = 2a + 1$

QUESTION 23

SHADE ONE BOX

What is the value of angle y in this diagram?

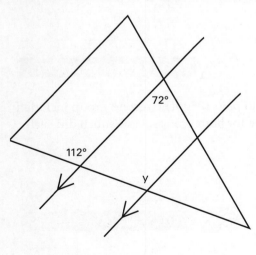

☐ 72° ☐ 68° ☐ 108° ☐ 112°

QUESTION 24

SHADE ONE BOX

Two normal dice are tossed. What is the probability of getting a total of 7?

☐ $\frac{5}{18}$ ☐ $\frac{1}{12}$ ☐ $\frac{5}{36}$ ☐ $\frac{1}{6}$

QUESTION 25

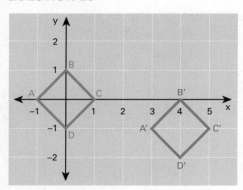

The blue shape in the diagram has been translated to the grey shape in the diagram by moving

☐ 4 horizontally and 2 vertically

☐ 1 horizontally and 4 vertically

☐ 4 horizontally and 2 vertically

☐ 4 horizontally and 1 vertically

QUESTION 26

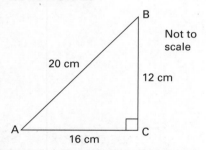

What is the area of this triangle?

☐ 96 cm²   ☐ 120 cm²   ☐ 160 cm²   ☐ 72 cm²

QUESTION 27

 WRITE YOUR OWN ANSWER

Two families booked plane tickets for a holiday in North
Queensland. The first group included 1 adult and 4 children and paid $950. The second group included
2 adults and 3 children and paid $975. The prices were the same for both groups. How much did each
child's ticket cost?

QUESTION 28

 WRITE YOUR OWN ANSWER

Write 2010 as a product of its prime factors.

QUESTION 29

 WRITE YOUR OWN ANSWER

A kitchen floor is to be tiled. The area to be tiled is 3.5 m wide by
4 m long. The tiles to be used are 20 cm × 35 cm. Each tile costs $14.50.
What is the total cost of the tiles?

QUESTION 30

When a boat docks at a wharf, a rope is used to tie it to a pylon. One-quarter of the pylon is sunk into the river bed and $\frac{2}{5}$ of the pylon is under water. What fraction of the pylon is above the water?

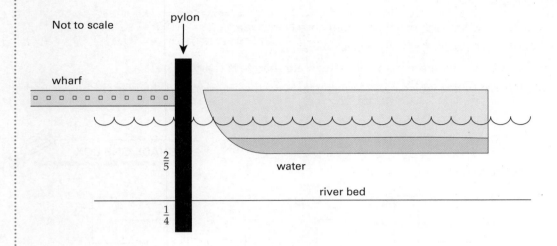

pylon

wharf

$\frac{2}{5}$

water

river bed

$\frac{1}{4}$

QUESTION 31

SHADE ONE BOX

The volume of this rectangular prism is 192 cm³.

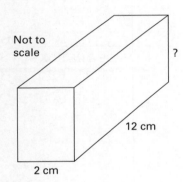

Not to scale

?

12 cm

2 cm

What is the length of this rectangular prism?

☐ 6 cm   ☐ 8 cm   ☐ 12 cm   ☐ 14 cm

QUESTION 32

 WRITE YOUR OWN ANSWER

The two triangles in the diagram below are similar.

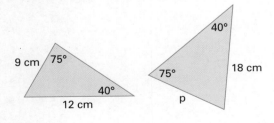

9 cm   75°

40°

75°

18 cm

40°

12 cm

p

What is the value of p? ☐ cm

Instructions

- A correct answer scores 1 mark, and an incorrect answer scores 0.
- Marks are not deducted for incorrect answers.
- No marks are given if more than one answer alternative is shaded.
- Choose the alternative which most correctly answers the question and shade in the box next to it.

QUESTION 1

SHADE ONE BOX

Which of these nets can be folded to make a cube?

☐   ☐   ☐   ☐

QUESTION 2

WRITE YOUR OWN ANSWER

A number is multiplied by itself and then 13 is added. The answer is 22. What is the number?

QUESTION 3

SHADE ONE BOX

The area of this triangle is 90 cm². What is the length of its base?

20 cm

☐ 4.5 cm   ☐ 9 cm   ☐ 18 cm   ☐ 20 cm

QUESTION 4

SHADE ONE BOX

What is the best estimate for $17 \times 32 \times 35 \times 9$?

☐ $10 \times 30 \times 30 \times 10$

☐ $10 \times 30 \times 40 \times 10$

☐ $20 \times 30 \times 40 \times 10$

☐ $20 \times 20 \times 30 \times 10$

QUESTION 5

WRITE YOUR OWN ANSWER

George cuts his birthday cake into 24 equal slices. He eats 25 per cent of the slices. How many slices are left?

QUESTION 6

SHADE ONE BOX

What is the hypotenuse of the following triangle PQR?

☐ PQ ☐ PR ☐ QR ☐ QP

QUESTION 7

WRITE YOUR OWN ANSWER

A tap leaks at a rate of 30 mL every 20 seconds.
How much water is wasted in 5 minutes?

mL

QUESTION 8

SHADE ONE BOX

If $y = -3x + 4$ and $x = -2$, what is the value of y?

☐ 1 ☐ -1 ☐ 2 ☐ 10

QUESTION 9

SHADE ONE BOX

A bee flies at a speed of 0.3 m/s. How long will it take to fly 60 m?

☐ 18 seconds ☐ 180 seconds ☐ 200 seconds ☐ 20 seconds

QUESTION 10

WRITE YOUR OWN ANSWER

Wendy saved $617.00 by buying her lounge suite at a 20% off sale.
What was the original price?

QUESTION 11

SHADE ONE BOX

A cube has side lengths of 20 cm. What is its volume in cubic metres?

☐ 0.008 ☐ 0.08 ☐ 0.8 ☐ 8

QUESTION 12

SHADE ONE BOX

There are 100 marbles in a bag. Twenty marbles are green and the rest are blue. Petra picks a marble without looking. What is her chance of picking a blue marble?

☐ 2:1 ☐ 4:5 ☐ 1:2 ☐ 5:4

QUESTION 13

SHADE ONE BOX

PQRS is a square.

What is the value of x?

- [ ] -1
- [ ] 1
- [ ] 2
- [ ] -2

QUESTION 14

SHADE ONE BOX

Which of these points is a point of intersection of these graphs?

- [ ] (3, 1)
- [ ] (0, -2)
- [ ] (1, 3)
- [ ] (0, 2)

QUESTION 15

WRITE YOUR OWN ANSWER

Most cookbooks provide a formula for calculating cooking times for different meats. When cooking roast turkey, one book suggests this formula:

$T = 50W + 25$ where T represents the cooking time (in minutes) and W represents the weight of the piece of the turkey (in kilograms).

What weight of turkey would cook in one hour?

kg

QUESTION 16

SHADE ONE BOX

What is the answer to $9.9 \div 0.3$?

- [ ] 3.3
- [ ] 0.033
- [ ] 0.33
- [ ] 33

9780170470674

QUESTION 17

SHADE ONE BOX

Four children have different pets. Using the clues below, work out who had the fish.

Bob's pet can't fly.

Cal's pet has hair, so does Deb's.

Deb's pet doesn't bark.

Use the table below to work out your answer.

	Bird	Cat	Dog	Fish
Amy				
Bob				
Cal				
Deb				

☐ Amy   ☐ Bob   ☐ Cal   ☐ Deb

QUESTION 18

SHADE ONE BOX

Which set shows values equivalent to 73 out of 125?

☐ 58%, $\frac{73}{125}$, 0.584

☐ $\frac{584}{1000}$, 5.84%, 0.0584

☐ 0.584, $\frac{292}{500}$, 58.4%

☐ 58.4%, $\frac{73}{125}$, 5.84

QUESTION 19

SHADE ONE BOX

In standard form, 0.00065 = ?

☐ 0.65×10^{-3}

☐ 6.5×10^{-4}

☐ 6.5×10^{-3}

☐ 6.5×10^{3}

QUESTION 20

WRITE YOUR OWN ANSWER

Roger buys seven postage stamps which cost a total of $3.85.
What is the cost of five of these postage stamps?

QUESTION 21

WRITE YOUR OWN ANSWER

$56 \times 84 = ?$

 WRITE YOUR OWN ANSWER

The stem-and-leaf plot below shows data about reaction time of left- and right-handed students.

									Stem									
									0.9									
								7	0.8									
									0.7									
								9	0.6									
						7	1	0	0.5	6								
						2	2	1	0.4	0	1	1	1	3				
9	7	6	6	6	5	4	2	1	0.3	1	1	2	2	2	4	5	9	
							8	6	0.2	3	3	6	4	7	8	9	9	
									0.1	4	7	7						
									0.0									

Mean: 0.43 sec
Median: 0.37 sec

Mean: 0.32 sec
Median: 0.31 sec

Left-handed | Right-handed

True or false? This data suggests that right-handed students have a faster reaction time than left-handed students.

 WRITE YOUR OWN ANSWER

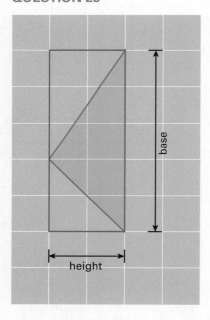

base

height

What is the area of this triangle? Each grid is 1 cm by 1 cm.

QUESTION 24

Which of the four arrangements of geometric solids matches this contour map?

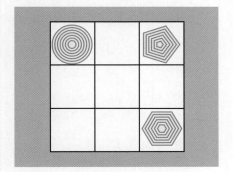

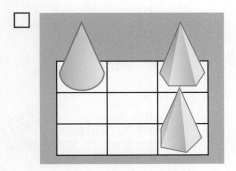

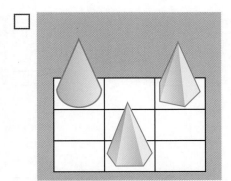

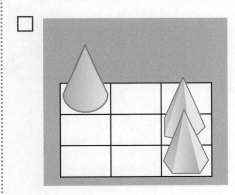

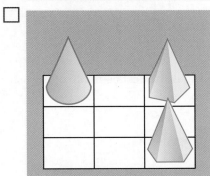

QUESTION 25

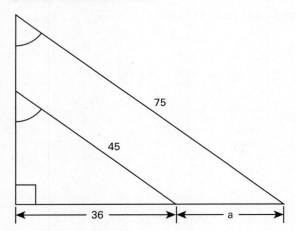

What is the value of a?

☐ 64   ☐ 60   ☐ 24   ☐ 34

QUESTION 26

$1^2 + 2^3 + 3^4 = ?$

☐ 20   ☐ 36   ☐ 90   ☐ 324

QUESTION 27

Which of the following is the best estimate of the size of the true bearing shown in the diagram?

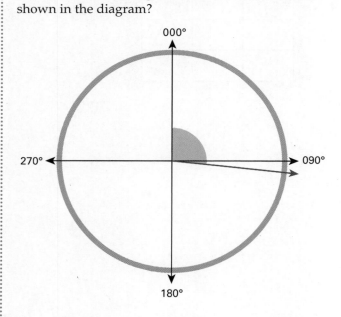

☐ 010°   ☐ 095°   ☐ 085°   ☐ 005°

QUESTION 28

SHADE ONE BOX

Which of the following statements best describes the shaded area in this Venn diagram?

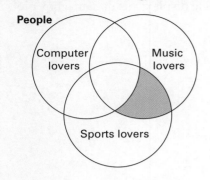

☐ people who love sports and music

☐ people who love music and computers

☐ people who love computers and sports

☐ people who love computers and music

QUESTION 29

WRITE YOUR OWN ANSWER

A packet of eight fruit buns costs $4.70. A packet of 12 fruit buns costs
$6.30. Leila needs to buy 44 fruit buns. What is the least amount she could pay for the fruit buns?

QUESTION 30

WRITE YOUR OWN ANSWER

A 10 m by 18 m function room is to be used by the Year 9 social
committee for a disco. A portable dance floor is to be installed to allow 6 m by 8 m of dance space.
What percentage of the floor space is set aside for dancing? Write your answer to include a fraction.

%

QUESTION 31

SHADE ONE BOX

The graph of $y = x^2 - 2$ is most like which of the following graphs?

☐ 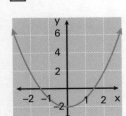   ☐ 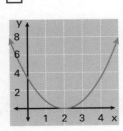   ☐ 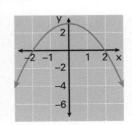   ☐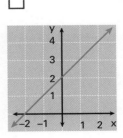

QUESTION 32

A quality-control engineer takes a random sample of 100 boxes of matches to check the claim that there are between 40 and 60 matches in each box. He displays his data in a box plot.

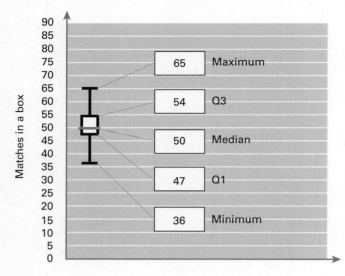

Which of the following statements is true about this box plot?

☐ Half of the boxes have between 47 and 54 matches.

☐ No box holds less than 47 matches.

☐ The mean number of matches in a box is 50.

☐ Three-quarters of the match boxes hold less than 65 matches.

9780170470674

Test 6: Calculator allowed

Instructions

- A correct answer scores 1 mark, and an incorrect answer scores 0.
- Marks are not deducted for incorrect answers.
- No marks are given if more than one answer alternative is shaded.
- Choose the alternative which most correctly answers the question and shade in the box next to it.

QUESTION 1

SHADE ONE BOX

If a = 6, what is the value of 9a?

☐ 96 ☐ 72 ☐ 54 ☐ 48

QUESTION 2

SHADE ONE BOX

At 7.00 a.m. the temperature in Brownlea was 10.8 °C. At midday it was
6.9 °C warmer. By 5.00 p.m. the temperature had dropped 7.3 °C from its midday temperature.
What was the temperature at 5.00 p.m.?

☐ 3.4 °C ☐ 10.4 °C ☐ 11.2 °C ☐ 25.0 °C

QUESTION 3

SHADE ONE BOX

Which one of these is an isosceles triangle?

☐ ☐ ☐ ☐

QUESTION 4

SHADE ONE BOX

Arrange in ascending order: 0.2, 20, $\frac{1}{2}$, 2%.

☐ 20, $\frac{1}{2}$, 0.2, 2%

☐ 2%, $\frac{1}{2}$, 20, 0.2

☐ 0.2, 2%, $\frac{1}{2}$, 20

☐ 2%, 0.2, $\frac{1}{2}$, 20

QUESTION 5

WRITE YOUR OWN ANSWER

If the sun rises at 6:27 a.m. and sets at 5:42 p.m., how many
hours and minutes are there from sunrise to sunset?

QUESTION 6

This is a rectangular prism.

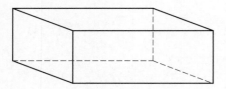

Which diagram shows the net of a rectangular prism?

☐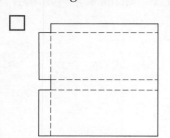

☐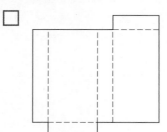

☐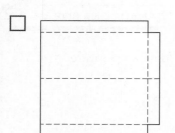

☐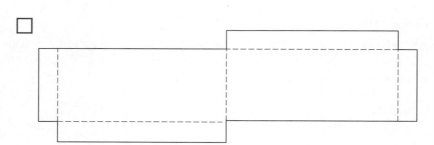

QUESTION 7

WRITE YOUR OWN ANSWER

The amount of water from a rainwater tank used by Bernadette's family for a period of four months is shown in the table below.

Month	1	2	3	4
Amount of water used (in litres)	6403	3712	5288	4957

What is the average amount of water used per month?

☐ litres

QUESTION 8

SHADE ONE BOX

A table of values is shown for x and y.

x	0	0.5	1	1.5	2
y	0	0.25	1	2.25	4

Which one of the following is the correct rule for y in terms of x?

☐ $y = 2x$ ☐ $y = x^2$ ☐ $y = \frac{x}{2}$ ☐ $y = 2x - 2$

QUESTION 9

SHADE ONE BOX

Fai bought a packet of eight pens for $6. What is the price of each pen?

☐ $0.60   ☐ $0.75   ☐ $1.33   ☐ $2

QUESTION 10

A flow chart is shown below.

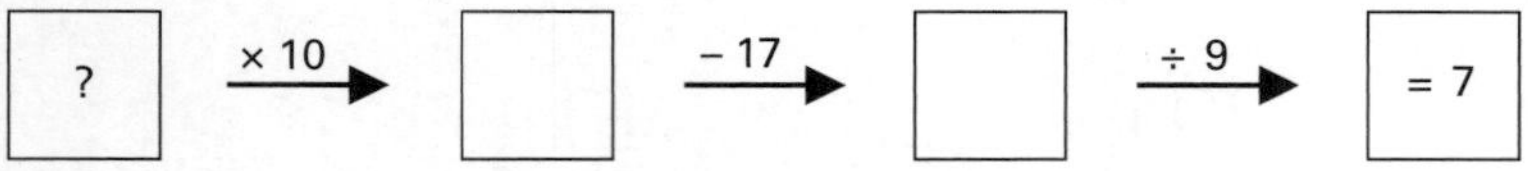

$$\boxed{?} \xrightarrow{\times 10} \boxed{} \xrightarrow{-17} \boxed{} \xrightarrow{\div 9} \boxed{= 7}$$

What is the starting number? □

QUESTION 11

This is a map of a tourist area.

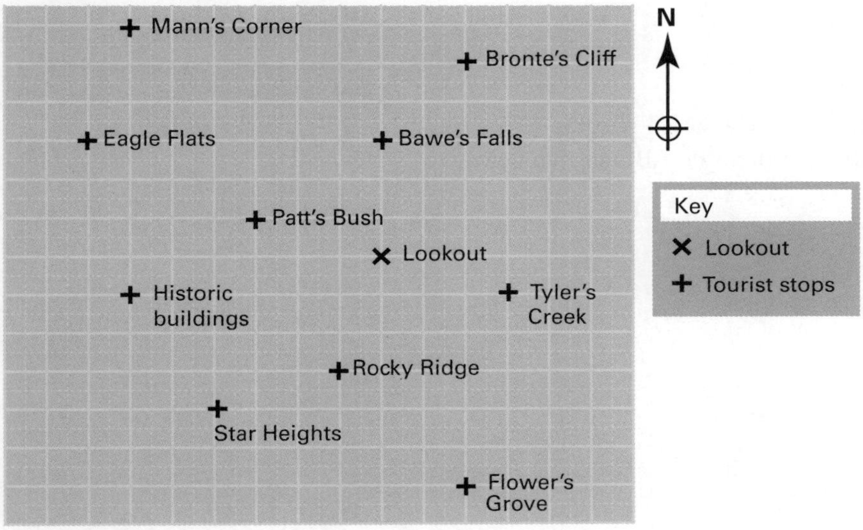

Amy is at the Lookout facing east. She turns 135° in an anticlockwise direction.
Which tourist attraction is she now facing?

☐ Mann's Corner ☐ Patt's Bush ☐ Star Heights ☐ Rocky Ridge

QUESTION 12

Rekha wants to enlarge this photograph of her pet cat.

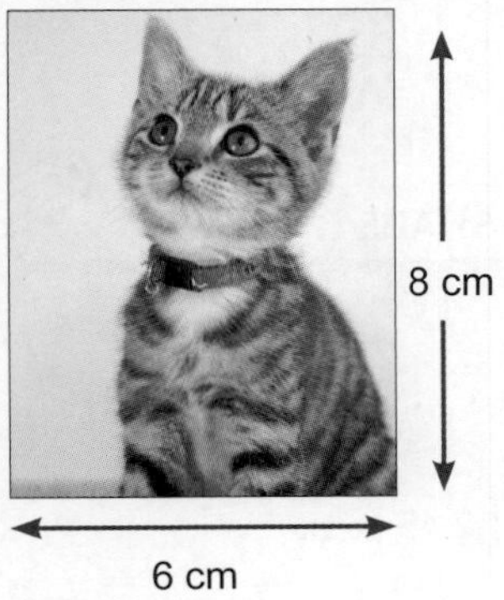

8 cm

6 cm

The enlarged photograph is four times as long and four times as wide as the original. What is the area of the enlarged photograph?

☐ 4 times the area of the original

☐ 8 times the area of the original

☐ 16 times the area of the original

☐ 24 times the area of the original

A recipe requires $\frac{2}{3}$ of a cup of flour. Ben is going to make double the recipe.
How many cups of flour does he need?

- [ ] $\frac{4}{6}$
- [ ] $\frac{3}{2}$
- [ ] $1\frac{1}{6}$
- [ ] $1\frac{1}{3}$

QUESTION 14

There are 40 marbles in a box. Fifteen marbles are green and the rest are white.
David picks a marble from the box without looking. What is the chance that David picks a white marble?

- [ ] 2 in 8
- [ ] 3 in 8
- [ ] 4 in 8
- [ ] 5 in 8

QUESTION 15

An equilateral triangle has a common side with a regular
pentagon on side BD. What is the size of angle ABC shown below?

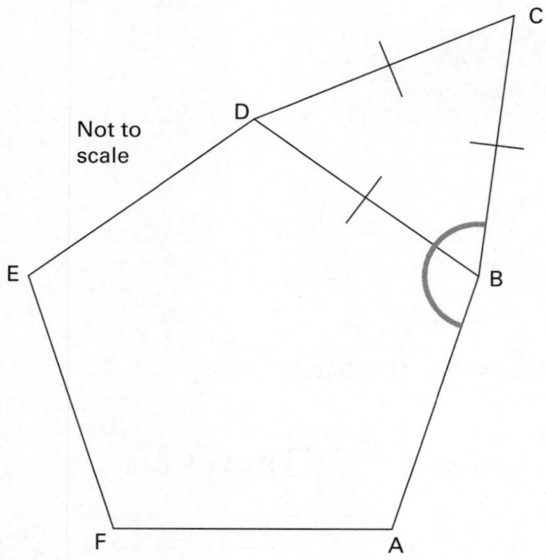

Not to scale

QUESTION 16

Melbourne is two hours behind New Zealand in time. A plane leaves New
Zealand at 10.45 a.m. and flies to Melbourne. The flight takes three hours.
What time is it in Melbourne when the plane arrives?

- [ ] 11.45 a.m.
- [ ] 12.45 p.m.
- [ ] 1.45 p.m.
- [ ] 2.45 p.m.

QUESTION 17

SHADE ONE BOX

In a three-hour period one morning, Johan spent time on the following activities.

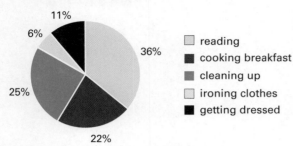

Johan's morning activities

- reading
- cooking breakfast
- cleaning up
- ironing clothes
- getting dressed

How much time did Johan spend cleaning up?

☐ 25 minutes

☐ 40 minutes

☐ 45 minutes

☐ 1 hour

QUESTION 18

SHADE ONE BOX

The object below was made using identical cubes

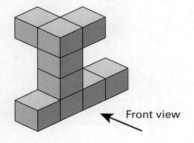

← Front view

Which diagram below shows the top view of this solid?

  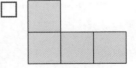 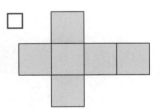

QUESTION 19

SHADE ONE BOX

A Christmas bauble is the shape of a sphere. It has a radius of 4 cm

The volume, V, of the sphere is given by the formula $V = \frac{4}{3}\pi r^3$, where r is the radius of the sphere in centimetres. What is the volume of this Christmas bauble closest to?

☐ 50 cm³   ☐ 120 cm³   ☐ 200 cm³   ☐ 270 cm³

QUESTION 20

SHADE ONE BOX

What is the value of x?

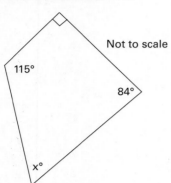

Not to scale

115°

84°

x°

☐ 71   ☐ 85   ☐ 96   ☐ 101

QUESTION 21

SHADE ONE BOX

Sujay takes one ball out of his bag without looking. Sujay chooses the bag that is most likely to choose a black ball. Which bag does he choose?

☐    ☐    ☐ 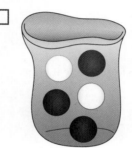   ☐

QUESTION 22

SHADE ONE BOX

Lynton buys watermelon at the supermarket. He constructed a line graph to show the price per kilogram for watermelon over a six-week period.

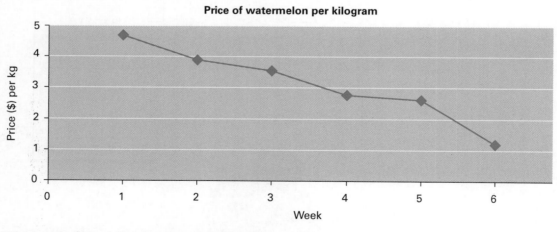

Price of watermelon per kilogram

Which one of the following statements is true?

☐ In week 1, the cost of 2 kg of watermelon was $8.

☐ The price of 1 kg of watermelon was less than $3 in week 4.

☐ The cost of 1 kg of watermelon halved in week 2.

☐ The largest drop in price for 1 kg of watermelon was from week 4 to week 5

9780170470674

QUESTION 23

Year 9 students attend a sports training day. Boys and girls attend the training day in the ratio 2:3.
If 60 girls attended the day, how many boys were there?

☐ 40   ☐ 24   ☐ 36   ☐ 20

QUESTION 24

This is a map of the route Sienna takes to walk to school. The map is drawn to scale.

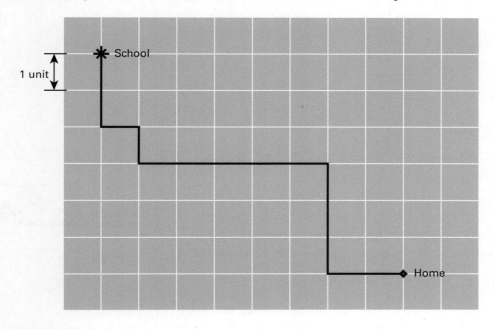

The total distance she travels to school is 840 metres. What is the scale of the map?

☐ 1 unit = 30 m

☐ 1 unit = 45 m

☐ 1 unit = 55 m

☐ 1 unit = 60 m

QUESTION 25

Sarah wants to draw the graph of the line $y = 3x – 4$. Which pair of points
would this line pass through?

☐ $(3, 0)$ and $(0, 4)$

☐ $(1\frac{1}{3}, 0)$ and $(0, -4)$

☐ $(0, 1\frac{1}{3})$ and $(4, 0)$

☐ $(0, 3)$ and $(-4, 0)$

QUESTION 26

SHADE ONE BOX

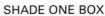

Jake is 1.8 m tall. He casts a shadow 3 m long when he stands 11 m from the base of a flagpole.

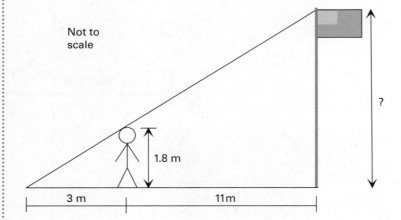

Not to scale

1.8 m

?

3 m

11m

What is the height of the flagpole?

☐ 6.6 m   ☐ 7.2 m   ☐ 8.4 m   ☐ 9.8 m

QUESTION 27

SHADE ONE BOX

Alexandra can ride her bike at a speed of 4 metres per second. How long, in hours and minutes, will it take her to ride 25 km to the nearest minute?

☐ 6 hours 15 minutes

☐ 6 hours 45 minutes

☐ 1 hour 44 minutes

☐ 1 hour 74 minutes

QUESTION 28

WRITE YOUR OWN ANSWER

What is the value of y in the equation $7y - 11 = 3y + 17$?

QUESTION 29

SHADE ONE BOX

This list shows the number of books read by members of a book club in March.

Number of books read	1, 2, 3, 5, 5, 6, 8, 8, 8, 9

Which one of the following is true for this data?

☐ mean < median < mode

☐ mean = median < mode

☐ mean < median = mode

☐ mean = median = mode

QUESTION 30

A bus stop shelter is shown in the diagram below. It is open at both ends.

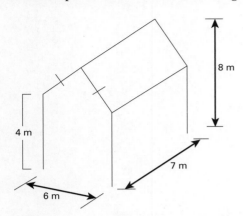

What is the area of metal needed to make the roof of this bus stop shelter?

☐ 98 m²   ☐ 84 m²   ☐ 70 m²   ☐ 56 m²

QUESTION 31

WRITE YOUR OWN ANSWER

Tahlia has read one-third of a book. If she reads another 65 pages,
she will have read half of the book. How many pages are there in this book?

☐ pages

QUESTION 32

WRITE YOUR OWN ANSWER

Ella uses these two conversion graphs.

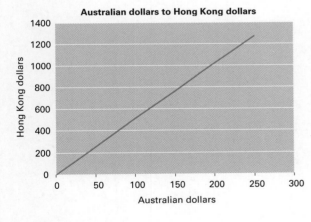

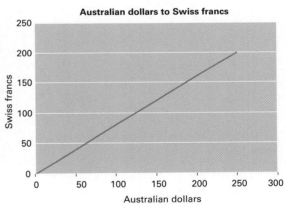

Approximately how many Swiss francs are equal in value to 1000 Hong Kong dollars?

☐ Swiss francs

Year 9 Numeracy
Non-calculator
Full-length Test 7

Writing time: 40 minutes

Use 2B pencil only

Instructions

· Write your **student name** in the space provided.
· You must be silent during the test.
· If you need to speak to the teacher, raise your hand. Do not speak to other students.
· Answer all questions using a 2B pencil.
· If you wish to change your answer, erase it very thoroughly and then write your new answer.
· Students are NOT permitted to bring a calculator into the test room.

Student name:

QUESTION 1

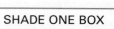

What is 7.08 kilolitres equivalent to?

SHADE ONE BOX

☐ 70.8 L

☐ 708 L

☐ 7080 L

☐ 70 800 L

QUESTION 2

SHADE ONE BOX

$7 + 2y - 5 - 9y = ?$

☐ -2 – 7y   ☐ -2 – 11y   ☐ 2 – 11y   ☐ 2 – 7y

QUESTION 3

SHADE ONE BOX

Which expression is equal to $2^3 \times 4^2$?

☐ $2 \times 3 \times 4 \times 2$

☐ $2 \times 2 \times 2 \times 3 \times 4 \times 2$

☐ $2 \times 2 \times 2 \times 2 \times 2 \times 2 \times 4 \times 4$

☐ $2 \times 2 \times 2 \times 2 \times 2 \times 2 \times 2$

QUESTION 4

SHADE ONE BOX

Which one of these is a right-angled isosceles triangle?

☐   ☐   ☐   ☐

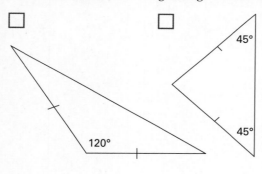 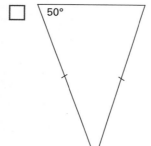 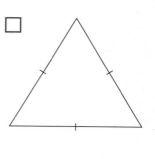

QUESTION 5

SHADE ONE BOX

What fraction has the same value as $3\frac{4}{5}$?

☐ $\frac{12}{5}$ ☐ $\frac{17}{5}$ ☐ $\frac{18}{5}$ ☐ $\frac{19}{5}$

QUESTION 6

A rectangle has an area of 144 cm².

Not to scale

?

8 cm

What is the length of the rectangle?

cm

QUESTION 7

What is the value of $5p^2$ when $p = -1$?

☐ -15 ☐ -5 ☐ 5 ☐ 15

QUESTION 8

How many edges does this triangular prism have?

edges

QUESTION 9

The graph below shows the preferred beverage of Year 11 students at a certain school.

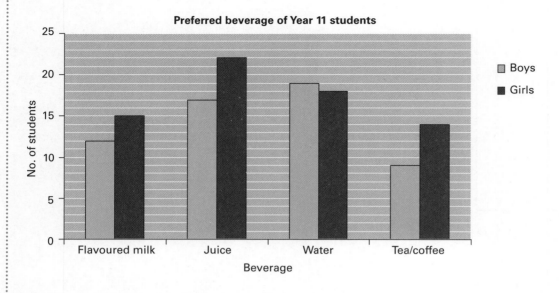

How many Year 11 students prefer water?

[   ]

QUESTION 10

SHADE ONE BOX

Which one of the following expressions is equivalent to 9x + 6?

☐ 2(3x + 2)

☐ 3(2x + 3)

☐ 3(3x + 2)

☐ 9(x + 6)

QUESTION 11

SHADE ONE BOX

What is the probability of rolling a number greater than 4 on a standard six-sided dice?

☐ $\frac{1}{3}$ ☐ $\frac{1}{2}$ ☐ $\frac{1}{4}$ ☐ $\frac{3}{8}$

QUESTION 12

SHADE ONE BOX

Which is the best estimate for 17 + 46 × 72 − 39?

☐ 10 + 40 × 80 − 30

☐ 20 + 50 × 80 − 40

☐ 10 + 40 × 70 − 30

☐ 20 + 50 × 70 − 40

9780170470674

QUESTION 13

This pie chart shows the types of messages received by Poppy's Florist in a single day.

Types of messages received by Poppy's Florist

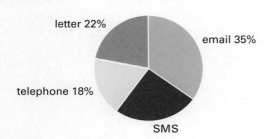

letter 22%

email 35%

telephone 18%

SMS

What percentage of the messages to Poppy's Florist were SMS messages? [ ] %

QUESTION 14

A number is added to itself and then 8 is added. The result is divided by 3. The answer is 10. What is the number?

[ ]

QUESTION 15

The object below is made from 8 cubes.

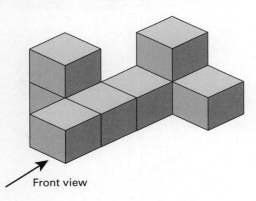

Front view

Which one of these shows the front view?

☐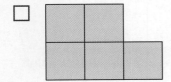

☐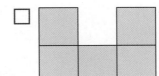

☐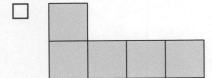

☐

QUESTION 16

Octavia folds this net to make a cube.

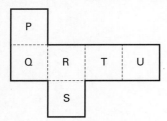

When Octavia constructs the cube, which face is opposite R?

☐ P ☐ Q ☐ S ☐ U

QUESTION 17

Solve for x: $6x + 1 = 4x + 5$

☐ $x = 2$

☐ $x = 3$

☐ $x = \frac{3}{5}$

☐ $x = 1\frac{2}{3}$

QUESTION 18

$\frac{5}{6} + \frac{2}{3} = ?$

☐ $\frac{7}{6}$ ☐ $\frac{7}{9}$ ☐ $1\frac{1}{2}$ ☐ $1\frac{1}{3}$

QUESTION 19

A meal at a restaurant costs $62. A goods and services tax of 10 per cent is added to the price. Which calculation will give the new price of the meal?

☐ $62 + 1.1$

☐ 62×1.1

☐ 62×0.1

☐ $62 + 0.1$

QUESTION 20

The data in the stem-and-leaf plot below shows the number of drinks sold per hour at a café on a single day.

Stem	Leaf
0	6 8
1	5 5 9
2	4 5 6 6 6
3	1 7

What is the mode of this data? ☐

QUESTION 21

The line graphs below show the average sunrise and sunset times each month in Sydney over a twelve-month period.

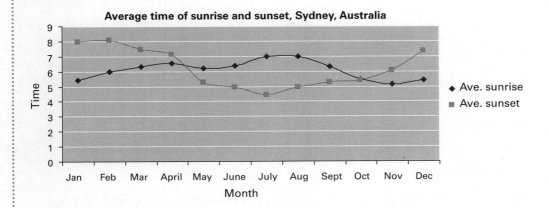

Average time of sunrise and sunset, Sydney, Australia

◆ Ave. sunrise
■ Ave. sunset

Which of the following statements is correct?

☐ In February, the average time when the sun rises is 6.00 a.m.

☐ The average time of sunrise is later than 6.00 a.m. for nine months of the year.

☐ The average time of sunset is before 6.00 p.m. from May to August only.

☐ In December, the average time when the sun sets is 5:30 a.m.

QUESTION 22

A horizontal cut is made through a cone, as shown below.

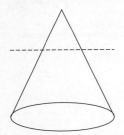

Which shape shows the cross-section made by the cut?

QUESTION 23

The coordinates of point X are (3, -6).

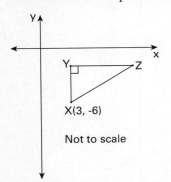

Not to scale

If XY ⊥ YZ, XY = 4 units and YZ = 5 units, what are the coordinates of point Z?

☐ (5, -2)   ☐ (2, -5)   ☐ (-2, 8)   ☐ (8, -2)

QUESTION 24

SHADE ONE BOX

What does $\sqrt{150}$ lie between?

- [ ] 10 and 12
- [ ] 11 and 13
- [ ] 13 and 15
- [ ] 50 and 80

QUESTION 25

SHADE ONE BOX

Phoebe has baggage to be loaded onto an aeroplane. Her bag weighs 30 kg.

Baggage is classified as overweight if it exceeds 23 kg. The cost of overweight baggage involves an airline handling fee of $40, plus $35 for each kilogram over the allowable weight.

What is the extra cost Phoebe is charged?

- [ ] $75
- [ ] $285
- [ ] $315
- [ ] $525

QUESTION 26

SHADE ONE BOX

A coin is tossed twice. What is the probability of getting a head and a tail in any order?

- [ ] $\frac{1}{8}$
- [ ] $\frac{1}{4}$
- [ ] $\frac{1}{3}$
- [ ] $\frac{1}{2}$

QUESTION 27

SHADE ONE BOX

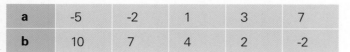

a	-5	-2	1	3	7
b	10	7	4	2	-2

The table shown above satisfies which equation?

- [ ] $b = a + 5$
- [ ] $b = a - 5$
- [ ] $b = -5 - a$
- [ ] $b = 5 - a$

QUESTION 28

SHADE ONE BOX

Which expression gives the total length of the dotted lines?

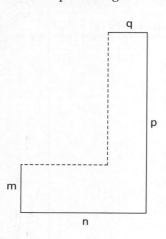

- [ ] $(m - p) + (n - q)$
- [ ] $(p - m) + (n - q)$
- [ ] $(p - m) - (n - q)$
- [ ] $(m + n) - (p + q)$

9780170470674

QUESTION 29

In how many ways is it possible to travel from Town A to Town B?

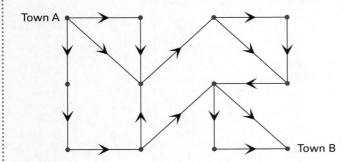

☐ 10   ☐ 12   ☐ 14   ☐ 17

QUESTION 30

WRITE YOUR OWN ANSWER

In the diagram below, AB∥CD.

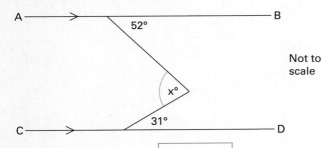

Not to scale

What is the value of x? ☐ degrees

QUESTION 31

WRITE YOUR OWN ANSWER

This diagram shows the course for a 70 km bicycle race. The start and finish are at the beach.

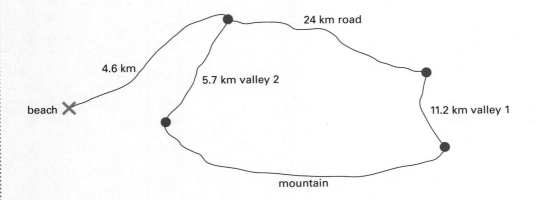

How far is the mountain section?

☐ km

A farmer is fencing a rectangular paddock using posts and wire.
He must put a post in every corner of the paddock and the posts must be 2 m apart.

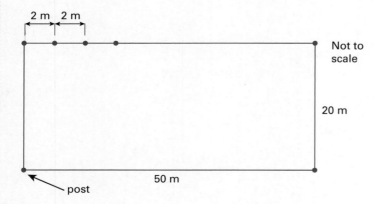

Not to scale

20 m

50 m

post

How many posts does the farmer need to fence this paddock?

posts

Year 9 Numeracy
Calculator allowed
Full-length Test 8

Writing time: 40 minutes

Use 2B pencil only

Instructions

- Write your **student name** in the space provided.
- You must be silent during the test.
- If you need to speak to the teacher, raise your hand. Do not speak to other students.
- Answer all questions using a 2B pencil.
- If you wish to change your answer, erase it very thoroughly and then write your new answer.
- Students are permitted to bring a calculator into the test room.

Student name:

QUESTION 1

SHADE ONE BOX

Two places are 5.4 cm apart on a map. The scale of the map is 1 cm represents 5 km.

What is the actual distance between the two places?

☐ 5.4 km   ☐ 54 km   ☐ 25 km   ☐ 27 km

QUESTION 2

WRITE YOUR OWN ANSWER

A recipe requires three-quarters of a cup of sugar. Alice is making double the amount. How much sugar does she need? Write your answer in simplest terms.

☐ cups

QUESTION 3

SHADE ONE BOX

What is the grid reference of the bird watch on this map?

☐ H8   ☐ I8   ☐ H7   ☐ I7

QUESTION 4

SHADE ONE BOX

Charlotte noticed that the temperature at her house was 11.5 °C when she woke up. Two hours later, the temperature had risen by 4.6 °C. Later in the day, a storm hit and the temperature fell by 5.3 °C. What was the temperature then?

☐ 10.8 °C   ☐ 9.8 °C   ☐ 11.8 °C   ☐ 18.3 °C

QUESTION 5

SHADE ONE BOX

What is the gradient of the line joining (-1, 2) and (3, 4)?

☐ $\frac{1}{2}$ ☐ $-\frac{1}{2}$ ☐ -2   ☐ 2

QUESTION 6

$1^3 + 2^3 + 3^3 = ?$

SHADE ONE BOX

☐ 18   ☐ 36   ☐ 84   ☐ 216

QUESTION 7

Which one of the following angles is equal to 72°?

SHADE ONE BOX

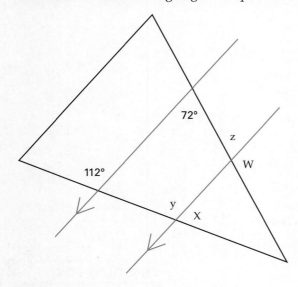

☐ W   ☐ X   ☐ y   ☐ z

QUESTION 8

What is the area of the triangle ABC shown?

SHADE ONE BOX

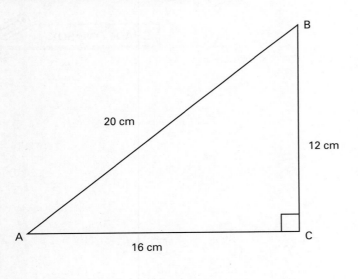

☐ 96 cm²   ☐ 120 cm²   ☐ 160 cm²   ☐ 72 cm²

QUESTION 9

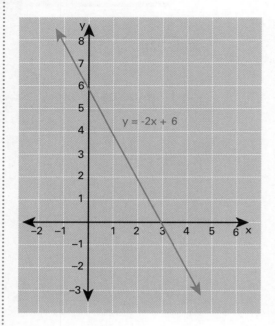

What are the coordinates of the x-intercept of the graph above? (,)

QUESTION 10

Which of the following is a net of a cube?

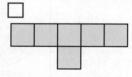 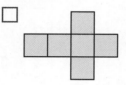 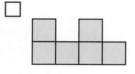 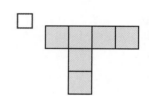

QUESTION 11

Look at the table below. The rule for this table is output = 3 × input + 5.

Input	Output
5	20
10	?
15	50
20	65

What is the missing value in the table?

☐ 45   ☐ 40   ☐ 35   ☐ 30

QUESTION 12

Which number is exactly halfway between $2\frac{1}{3}$ and $3\frac{2}{3}$?

QUESTION 13

SHADE ONE BOX

When the advertised new car price of $31 689 is reduced at the end of the financial year to $29 907, what is the approximate percentage decrease offered?

☐ 6% ☐ 17.82% ☐ 5.6% ☐ 5.9%

QUESTION 14

WRITE YOUR OWN ANSWER

The table shows the amount of time Tanya spent studying over five days.

Day	Time
Monday	25 min
Tuesday	40 min
Wednesday	48 min
Thursday	1 h
Friday	27 min

What was the average time Tanya spent studying?

☐ minutes

QUESTION 15

WRITE YOUR OWN ANSWER

One-sixth of Gemma's income is spent on train fares.
This represents $33. What is Gemma's total income?

☐

QUESTION 16

SHADE ONE BOX

What is the length of the hypotenuse of triangle ABC?

B

12 cm

A

16 cm

C

☐ 28 cm ☐ 20 cm ☐ 10.58 cm ☐ 13.9 cm

QUESTION 17

SHADE ONE BOX

The graph below shows the number of videos rented by some families from Top Flix in one week.

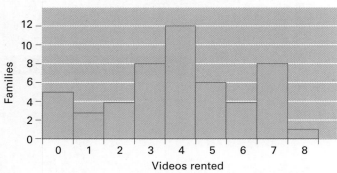

Video rentals from Top Flix

How many families rented more than four videos?

☐ 19   ☐ 12   ☐ 6   ☐ 31

QUESTION 18

SHADE ONE BOX

Ming bought a packet of 40 biscuits on Saturday. On Sunday she ate half of them. On Monday she ate one-quarter of the remaining biscuits. How many biscuits did Ming have left on Tuesday?

☐ 10   ☐ 15   ☐ 16   ☐ 20

QUESTION 19

 WRITE YOUR OWN ANSWER

Two normal dice are tossed. What is the probability of getting a total of 1?

QUESTION 20

SHADE ONE BOX

Which of the following curves could be described by $y = (x + 2)^2$?

☐ 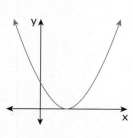   ☐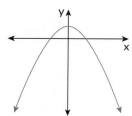

☐ 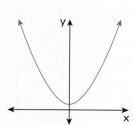   ☐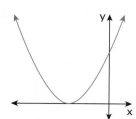

9780170470674

QUESTION 21

Maxwell gets pocket money for walking the dog. He is paid $4 for every hour, rounded to the nearest hour, that he walks each week. The table below shows his record for this week.

Monday	55 min
Tuesday	45 min
Wednesday	50 min
Thursday	20 min
Friday	1 h

How much pocket money did he get this week? $ _____

QUESTION 22

Which of the following shapes has exactly one axis of symmetry?

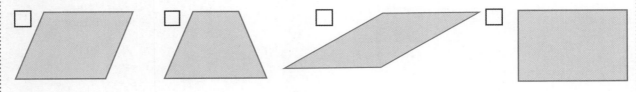

QUESTION 23

What is the height (h) of the shed shown in this diagram?
Write your answer correct to two decimal places.

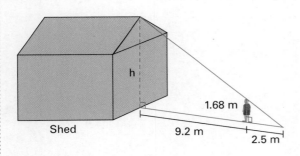

Shed   9.2 m   2.5 m   1.68 m

_____ m

QUESTION 24

A car salesperson is paid a base salary of $10 000 p.a. plus 4.25%
on the value of cars he sells.

How much does he earn in a year when he sells $1.44 million worth of cars? $ _____

QUESTION 25

A cube has a surface area of $6a^2$. If each side length is doubled,
what is the surface area of the new cube?

_____

QUESTION 26

Using the graph, what is the solution of $(x + 3)(3 - x) = 5$?

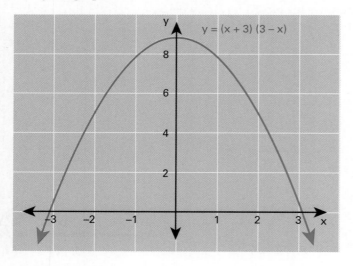

$y = (x + 3)(3 - x)$

☐ $x = 0$

☐ $x = -1.5$ or $x = 1.5$

☐ $x = -3$ or $x = 3$

☐ $x = -2$ or $x = 2$

QUESTION 27

WRITE YOUR OWN ANSWER

Using the map below, what is the true bearing of the windmill from the mud hut? Give your answer to the nearest 10 degrees.

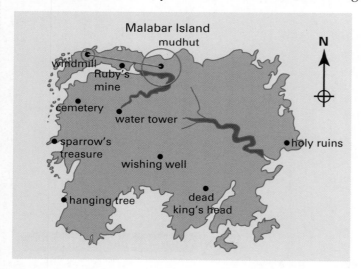

Malabar Island

mudhut

windmill

Ruby's mine

N

cemetery

water tower

sparrow's treasure

holy ruins

wishing well

hanging tree

dead king's head

☐ degrees

QUESTION 28

WRITE YOUR OWN ANSWER

The formula for calculating the volume of a cylinder is $V = \pi r^2 h$.
What is the height of a cylinder with a volume of 110 cm³ and radius of 2.5 cm?
Write your answer to two significant figures.

☐ cm

QUESTION 29

$(7.56 \times 10^{23}) \times (3.6 \times 10^{-6}) = ?$

- ☐ 2.7216×10^{18}
- ☐ 2.7216×10^{30}
- ☐ 2.1×10^{29}
- ☐ 2.1×10^{17}

QUESTION 30

 WRITE YOUR OWN ANSWER

Find the volume of water, in litres, that could be collected from rainwater run-off to fill each of these water tanks. (Remember: 1 L can fit in a cube with a side length of 10 cm.) Write you answer to the nearest litre.

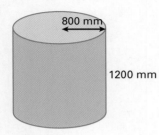

800 mm

1200 mm

☐ L

QUESTION 31

 **SHADE ONE BOX**

A coin is tossed four times. What is the probability of getting four heads?

- ☐ 0.25
- ☐ 0.125
- ☐ 0.625
- ☐ 0.0625

QUESTION 32

 WRITE YOUR OWN ANSWER

A school has 100 students in Year 9. They have three sports to choose from: softball, tennis or athletics. Some students do not play a sport.

- 5 students select all three sports.
- 13 select softball and tennis.
- 17 select tennis and athletics.
- 21 select athletics and softball.
- 46 select softball.
- 35 select tennis.
- 55 select athletics.

Use a Venn diagram or another method to determine the probability of a student selecting exactly one sport.

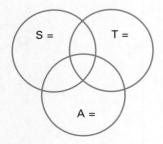

S =   T =

A =

☐

Year 9 Numeracy
Non-calculator
Full-length Test 9

Writing time: 40 minutes

Use 2B pencil only

Instructions

· Write your **student name** in the space provided.
· You must be silent during the test.
· If you need to speak to the teacher, raise your hand. Do not speak to other students.
· Answer all questions using a 2B pencil.
· If you wish to change your answer, erase it very thoroughly and then write your new answer.
· Students are NOT permitted to bring a calculator into the test room.

Student name:

QUESTION 1

SHADE ONE BOX

Stacey is making this pattern from straws.

 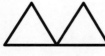 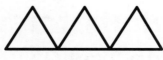 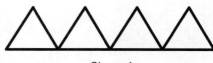

Shape 1   Shape 2   Shape 3   Shape 4

The table below shows the number of straws in each shape.

Shape	1	2	3	4	5
Number of straws	3	6	9	12	?

How many straws does Stacey need for Shape 5?

☐ 14   ☐ 15   ☐ 16   ☐ 20

QUESTION 2

SHADE ONE BOX

Which of the following sets of numbers is arranged in order from smallest to largest?

☐ 0.1101, 0.111, 0.11

☐ 0.1101, 0.11, 0.111

☐ 0.11, 0.1101, 0.111

☐ 0.11, 0.111, 0.1101

QUESTION 3

SHADE ONE BOX

What percentage of the diagram is shaded?

☐ 60%   ☐ 40%   ☐ 24%   ☐ 36%

QUESTION 4

WRITE YOUR OWN ANSWER

Write the perimeter of the shape below in the box. All measurements are in centimetres.

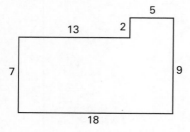

☐ cm

9780170470674

QUESTION 5

SHADE ONE BOX

The pie chart below shows a sample of 500 Year 9 students' favourite artistic activities.

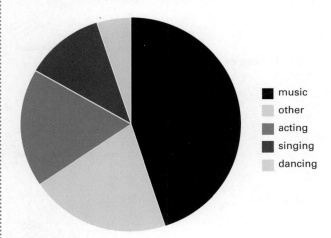

Favourite artistic activities

- ■ music
- ■ other
- ■ acting
- ■ singing
- ■ dancing

Which of the following conclusions can be drawn from the graph?

☐ More than one-fifth of the sample preferred dancing.

☐ More than one-quarter of the students surveyed preferred acting.

☐ Less than one-tenth of the students preferred singing.

☐ Less than half the students preferred music.

QUESTION 6

WRITE YOUR OWN ANSWER

What is the area of a triangle with base length 10 m and height 6 m?

 cm²

QUESTION 7

WRITE YOUR OWN ANSWER

Sam the concreter mixes one part cement to two parts water to
three parts gravel to make his concrete. He uses 6 kg of cement for this job. How much gravel does he need?

 kg

QUESTION 8

SHADE ONE BOX

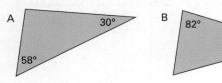 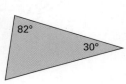 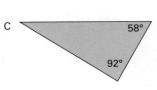

Which of the above triangles are similar?

☐ B and C   ☐ A and B   ☐ All three   ☐ A and C

QUESTION 9

SHADE ONE BOX

As a basic numeral, $2.51 \times 10^{-3} = ?$

☐ 2510 ☐ 0.00251 ☐ 0.0251 ☐ 25.1

QUESTION 10

SHADE ONE BOX

In a family of three children, what is the probability of there being exactly two girls?

☐ $\frac{1}{8}$ ☐ $\frac{1}{4}$ ☐ $\frac{3}{8}$ ☐ $\frac{1}{2}$

QUESTION 11

SHADE ONE BOX

What is $\sqrt{120}$ between?

☐ 8 and 9 ☐ 9 and 10 ☐ 10 and 11 ☐ 11 and 12

QUESTION 12

SHADE ONE BOX

Which expression is equivalent to $3 - 2x$?

☐ $-2x + 3$ ☐ $2x - 3$ ☐ $-3 + 2x$ ☐ $-3 - 2x$

QUESTION 13

SHADE ONE BOX

The following stem-and-leaf plot shows the result, in centimetres, of
measuring 23 boys aged 13 years. What is the median height of these boys to the nearest centimetre?

13 year-old boys' heights

19	
18	8
17	2 4
16	1 6
15	0 0 1 2 4 5 8 8
14	3 4 5 5 7 8 8
13	1 3 6
12	
11	
10	

☐ 145 cm ☐ 150 cm ☐ 146 cm ☐ 158 cm

QUESTION 14

WRITE YOUR OWN ANSWER

What is the point of intersection of these two graphs?

(　,　)

QUESTION 15

Which of the following describes the transformation needed to move the black object to the position of the blue object?

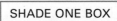

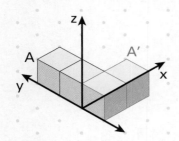

☐ 90° clockwise about the z-axis

☐ 90° anticlockwise about the x-axis

☐ 90° anticlockwise about the z-axis

☐ 90° about the line y = x

QUESTION 16

WRITE YOUR OWN ANSWER

If 3 is added to x and the result doubled, the answer is 16.
What is the value of x?

QUESTION 17

SHADE ONE BOX

Beryl has 24 green buttons and 48 red buttons. What fraction of the buttons is red?

☐ $\frac{1}{2}$ ☐ $\frac{1}{3}$ ☐ $\frac{3}{4}$ ☐ $\frac{2}{3}$

QUESTION 18

SHADE ONE BOX

The product of y and y – 7 can be written as?

☐ y + (y – 7)   ☐ y(y – 7)   ☐ $y^2 - 7$ ☐ 7y

QUESTION 19

SHADE ONE BOX

At 12 noon EST in Sydney, the times in some other international cities are as follows:

London 2:00 a.m.	Oslo 3:00 a.m.
Tokyo 11:00 a.m.	Cape Town 4:00 a.m.
Beijing 10:00 a.m.	Rangoon 8:30 a.m.
Mumbai 7:30 a.m.	Istanbul 4:00 a.m.

In which city is it 3.00 a.m. if it is 5.00 a.m. in Sydney?

☐ London

☐ Tokyo

☐ Beijing

☐ Cape Town

QUESTION 20

In the diagram below, which of the angles are equal to 63°?

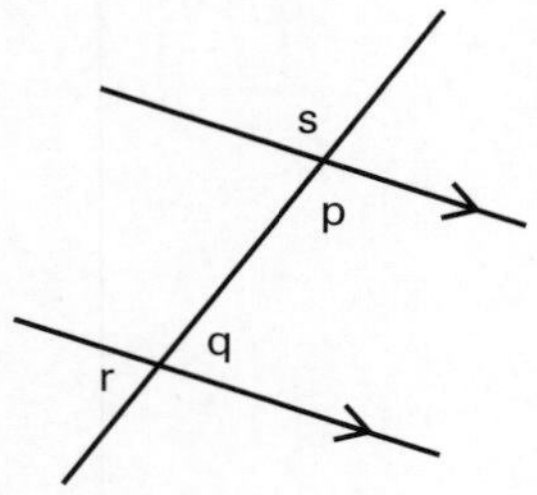

☐ p and s   ☐ p and q   ☐ q and r   ☐ s and q

QUESTION 21

WRITE YOUR OWN ANSWER

What is the area of the figure below? Write your answer to the nearest whole number.

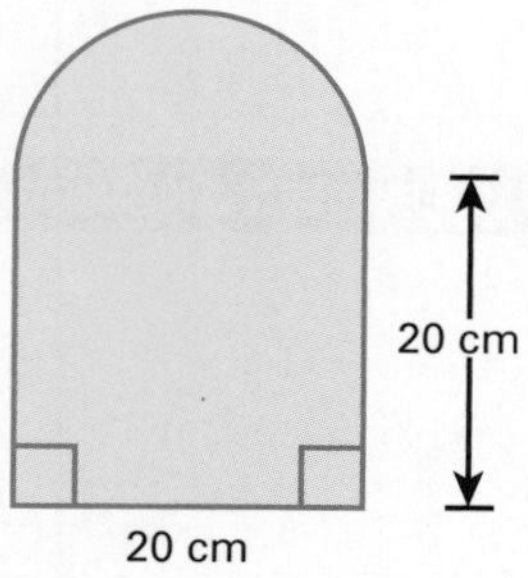

20 cm

20 cm

☐ cm²

QUESTION 22

SHADE ONE BOX

A bag contains three red, two yellow, one green and four blue marbles. What is the probability of picking a red marble at random?

☐ 3   ☐ $\frac{3}{7}$ ☐ 0.7   ☐ 0.3

QUESTION 23

SHADE ONE BOX

Shade the box that shows 5.03 correct to 2 significant figures.

☐ 5.0   ☐ 5.03   ☐ 5.1   ☐ 5

QUESTION 24

SHADE ONE BOX

Below is a set of 8 scores:

3, 5, 5, 10, 10, 13, 13, 50

What will change if score 50 is removed?

☐ mean only

☐ mean and median

☐ mean and mode

☐ mean and range

9780170470674

QUESTION 25

Two numbers added together equal -1. The same two numbers multiplied together equal -72.

What are the two numbers? ☐ and ☐

QUESTION 26

At the Boxing Day sales, a TV was advertised for $600, which was 40 per cent off the original price. What was the original price?

$ ☐

QUESTION 27

The sum of the ages of Sara and Tomo is 20, the sum of the ages of Sara and Ubi is 16, and the sum of the ages of Tomo and Ubi is 12. Find Ubi's age.

☐ years old

QUESTION 28

Which of the following is the table of values for $y = 6 - 2x^2$?

☐

x	-2	-1	0	1	2
y	10	8	6	4	2

☐

x	-2	-1	0	1	2
y	2	-4	-6	-4	2

☐

x	-2	-1	0	1	2
y	-2	4	6	4	-2

☐

x	-2	-1	0	1	2
y	14	8	6	8	14

QUESTION 29

Which number has the greatest value?

☐ $\frac{1}{3}$ ☐ $\sqrt{0.09}$ ☐ 0.31   ☐ $(0.3)^2$

QUESTION 30

Which of the following are the Pythagorean triads?

a   (3, 4, 5)
b   (5, 10, 15)
c   (7, 24, 25)

☐ a only   ☐ a and c   ☐ b and c   ☐ all three

QUESTION 31

A class is asked to suggest a way of randomly surveying people to find out their opinions about a general question. The suggestions are:

a everyone at a school has their name on a ticket. The tickets are put in a barrel and three are drawn out to be school representatives.

b people from the Adelaide White Pages phone directory are phoned at random in a survey.

c a survey of medium-sized Australian businesses is conducted by choosing every 200th business from GST-registered businesses.

The random methods are:

☐ all three ☐ a and b ☐ b and c ☐ a and c

QUESTION 32

Three diagrams of circuit designs are shown below.

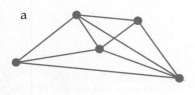 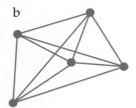 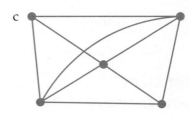

Which design(s) can be redrawn without wires crossing?

☐ none ☐ all ☐ a and c ☐ b and c

Year 9 Numeracy
Calculator allowed
Full-length Test 10

Writing time: 40 minutes

Use 2B pencil only

Instructions

- Write your **student name** in the space provided.
- You must be silent during the test.
- If you need to speak to the teacher, raise your hand. Do not speak to other students.
- Answer all questions using a 2B pencil.
- If you wish to change your answer, erase it very thoroughly and then write your new answer.
- Students are permitted to bring a calculator into the test room.

Student name:

QUESTION 1

SHADE ONE BOX

Which expression is equivalent to 6 – 5t?

☐ 5t – 6 ☐ 6t – 5 ☐ -6 + 5t ☐ -5t + 6

QUESTION 2

SHADE ONE BOX

What does $4^3 - 5^2 = ?$

☐ 39 ☐ 31 ☐ 2 ☐ -1

QUESTION 3

SHADE ONE BOX

Which of the following is the smallest number?

☐ $\frac{8}{5}$ ☐ $1\frac{1}{3}$ ☐ $\frac{11}{9}$ ☐ 1.4

QUESTION 4

WRITE YOUR OWN ANSWER

What is the size of the marked angle in this triangle?

QUESTION 5

SHADE ONE BOX

A chef at a restaurant makes an 8 L pot of soup to serve for entrees one evening. He dishes out the soup into bowls that hold 300 mL. How many full servings of soup can be served in his restaurant tonight?

☐ 26 ☐ 27 ☐ 37 ☐ 38

QUESTION 6

SHADE ONE BOX

A square has an area of 121 cm². What is the length of its diagonal closest to?

☐ 10 cm ☐ 11 cm ☐ 16 cm ☐ 18 cm

9780170470674

QUESTION 7

SHADE ONE BOX

A blue whale weighs 80 tonnes. It eats 4 tonnes of krill each day. What percentage of its own weight does it eat each day?

☐ 5%   ☐ 8%   ☐ 12%   ☐ 15%

QUESTION 8

SHADE ONE BOX

Anne drew this shape.

This shape is rotated clockwise 180° about point P. Which shape below does it now look like?

☐   ☐   ☐   ☐

QUESTION 9

SHADE ONE BOX

Sticks are used to make this pattern of triangles.

In this pattern, what is the rule for the number of sticks?

☐ 3 × number of triangles

☐ 2 × number of triangles

☐ 3 × number of triangles − 1

☐ 2 × number of triangles + 1

QUESTION 10

SHADE ONE BOX

A rule for q in terms of p is q = 5 − 2p. When p = 6.25, what is the value of q?

☐ -7.5   ☐ -3.25   ☐ 7.5   ☐ 9.25

QUESTION 11

SHADE ONE BOX

Which number is exactly halfway between $2\frac{1}{3}$ and $3\frac{1}{2}$?

☐ $2\frac{5}{6}$ ☐ $2\frac{2}{5}$ ☐ $2\frac{11}{12}$ ☐ $3\frac{1}{6}$

QUESTION 12

A hand-held fan is in the shape of a quadrant. The two straight sides are
12 cm long. The curved length from J to K is to have ribbon attached to it for decoration.

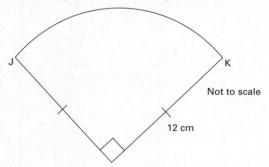

J

K

Not to scale

12 cm

What length is the curved length JK closest to?

☐ 19 cm ☐ 31 cm ☐ 43 cm ☐ 62 cm

QUESTION 13

 WRITE YOUR OWN ANSWER

The pie chart below shows the various milkshake flavours and
percentages sold at a café over a one week period in summer.

Milkshakes sold at café

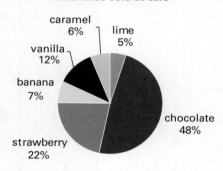

caramel
6%
lime
5%
vanilla
12%
banana
7%
chocolate
48%
strawberry
22%

If 250 milkshakes were sold altogether, how many vanilla
milkshakes were sold during this particular week?

vanilla milkshakes

QUESTION 14

What is the size of angle a°?

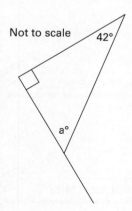

Not to scale 42°

a°

☐ 38° ☐ 48° ☐ 132° ☐ 142°

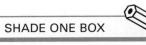
QUESTION 15

Neve spins this arrow 100 times.

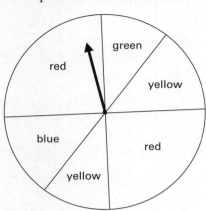

Which table is most likely to represent her results?

☐ ☐ ☐ ☐

Colour	No. of spins
Red	30
Blue	20
Yellow	25
Green	25

Colour	No. of spins
Red	40
Blue	30
Yellow	20
Green	10

Colour	No. of spins
Red	60
Blue	10
Yellow	20
Green	10

Colour	No. of spins
Red	30
Blue	25
Yellow	30
Green	15

QUESTION 16

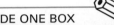
The most popular destinations for Australians to visit overseas in 2008–09 are shown below.

Country	Number of Australians
New Zealand	955 300
United States of America	500 000
Indonesia	436 000
United Kingdom	420 200
Thailand	378 400
China	268 000
Fiji	220 900
Singapore	213 700
Malaysia	205 200
Hong Kong	200 100

How many more people visited the United Kingdom than Singapore and Malaysia combined?

☐ 418 900

☐ 411 700

☐ 8500

☐ 1300

QUESTION 17

SHADE ONE BOX

This graph shows the number of children in each family for students in a Year 9 mathematics class.

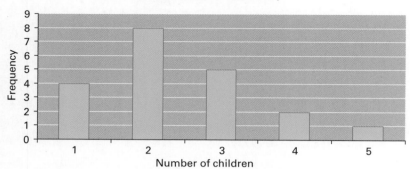

Number of children in each family

A new student who joins the class has two brothers. Which of the following statements is true?

☐ The mean increases and the mode does not change.

☐ The mean decreases and the mode does not change.

☐ The mode changes and the mean does not change.

☐ The mode and the mean both change.

QUESTION 18

 WRITE YOUR OWN ANSWER

Alisha works for $16.55 an hour. If she works for 18 hours in one week, how much will Alisha earn?

QUESTION 19

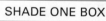 SHADE ONE BOX

What is the value of x?

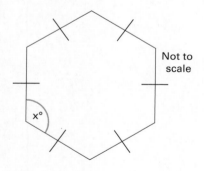

Not to scale

☐ 144°　　　　☐ 120°　　　　☐ 106°　　　　☐ 95°

QUESTION 20

 SHADE ONE BOX

The formula $F = \frac{9}{5}C + 32$ is used to convert degrees Celsius (°C) to degrees Fahrenheit (°F). When the temperature is 28 °C, what is the value of F?

☐ 82.4　　　　☐ 61.8　　　　☐ 50　　　　☐ 47.6

QUESTION 21

SHADE ONE BOX

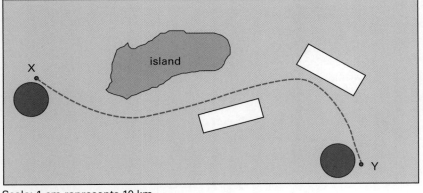

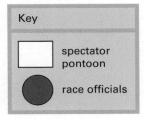

Key

spectator pontoon

race officials

Scale: 1 cm represents 10 km

The diagram above shows the layout for a waterskiing race. The distance from X to Y along a river is 12 cm. How long will it take a waterskier to complete the race at a speed of 150 km/h?

☐ 12 hours 50 minutes

☐ 12 hours 30 minutes

☐ 1 hour 15 minutes

☐ 48 minutes

QUESTION 22

SHADE ONE BOX

Which one of these points lies on the straight line that includes (3, 11) and (5, 15)?

☐ (1, 8)　　　　☐ (6, 17)　　　　☐ (10, 30)　　　　☐ (11, 35)

QUESTION 23

SHADE ONE BOX

A feature wall is constructed in a garden in this shape.

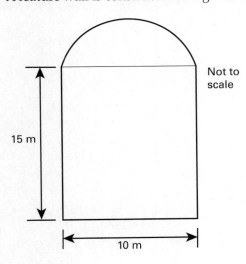

Not to scale

15 m

10 m

If it is to be painted, what is its area closest to?

☐ 189 m²　　　　☐ 229 m²　　　　☐ 307 m²　　　　☐ 464 m²

QUESTION 24

The net of a triangular prism is shown below.

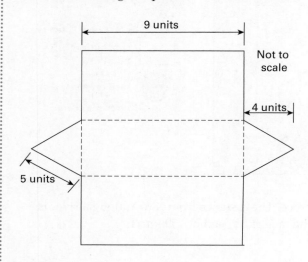

What is the volume of this solid?

☐ 90 cm³   ☐ 108 cm³   ☐ 168 cm³   ☐ 180 cm³

QUESTION 25

WRITE YOUR OWN ANSWER

Given a = 6 and b = -7, what is the value of a(a − b)?

QUESTION 26

SHADE ONE BOX

What is the length of x?

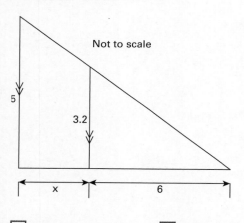

☐ 3.375   ☐ 9.375   ☐ 10.8   ☐ 15.625

QUESTION 27

SHADE ONE BOX

Here is the graph of a linear equation.

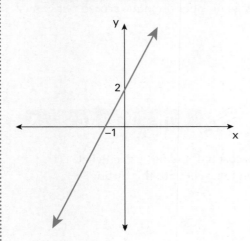

If this straight line is extended at both ends, which one of the following points will lie on this line?

☐ (-3, -8)   ☐ (-7, -12)   ☐ (5, 11)   ☐ (12, 27)

QUESTION 28

SHADE ONE BOX

$y = 2x + 3$

$y = 5x - 9$

Which value of x is the solution to both equations?

☐ x = 4   ☐ x = 2   ☐ x = -2   ☐ x = -4

QUESTION 29

 WRITE YOUR OWN ANSWER

One Australian dollar buys 0.52 British pounds. Using this
exchange rate, how many Australian dollars can be bought with 650 British pounds?

$ ☐

QUESTION 30

 WRITE YOUR OWN ANSWER

Red, blue and white discs are placed in a box in the ratio 2:5:3
respectively. There are 40 discs in the box. If 4 red discs and 6 white discs
are added to the box, what fraction of the discs in the box is now white?

☐

9780170470674

✏️ **WRITE YOUR OWN ANSWER**

After nine games, Patrick's average (mean) number of points scored per game was 12. After another seven games, his average had increased to 16. What was his average number of points scored per game overall?

| | points scored per game

✏️ **WRITE YOUR OWN ANSWER**

An airline is offering a special deal of buy one full-price ticket for $1350 and pay 70 per cent for the second ticket. Children's tickets cost half the full-price ticket. How much would it cost a family of two adults and three children to travel with this airline?

$ | |